THE ARCHANGELS

BOOK TWO, THE CIA AREA 51 CHRONICLES

The Complete Illustrated History of the CIA at Area 51

By

TD Barnes
"Thunder"

TD Barnes Copyright 2017

Contents
Preface
Introduction
Chapter 1 – Replacing the U-2
Chapter 2 – Electronic Warfare
Chapter 3 – Project 51
Chapter 4 – Area 51 Cadre
Chapter 5 – Working at Area 51
Chapter 6 – The Buildup
Chapter 7 – Article 123 Down
Chapter 8 – Targeting the OX
Chapter 9 – Mission Planning
Chapter 10 – Operation BLACK SHIELD
Chapter 11 – The USS Pueblo Incident2
Chapter 12 – BLACK SHIELD Missions
Chapter 13 – Mission Analysis
Epilogue
Bibliography
About the Author
Other Books by Author

Dedication

Primarily, I dedicate this book to all who served the CIA at its Groom Lake operating facility at Area 51 and the unknown participants who served surreptitiously at the affiliated power projection locations throughout the world.

Acknowledgments

My writing the three books, *The Angels* and *The Archangels,* and the *Company Business* of my *The CIA's Area 51 Chronicles* brings back memories of the pride and honor that my special projects team shared as we worked compartmentalized and in utmost secrecy to win the Cold War. Ask anyone with whom I served what the best era of their distinguished careers was and everyone proudly states that it was their time with boots on the ground at Area 51. Ask them what they remember most, and they brag on the food. Ask them whom they remember most, and they smile and without hesitation identify Murphy Green who ran the mess hall.

I especially acknowledge my family. For years, my faithful spouse, Doris, dropped me off in a secured area to catch a plane on Monday morning. Sometimes I flew out of Nellis AFB, and at other times at a highly-secured spot along Sunset Boulevard and just inside the perimeter fence of the McCarran Airport in Las Vegas. She picked me up when I returned Friday evening, not knowing where I went or what I did. She never asked. Finally, I can tell her.

I humbly acknowledge my Area 51 CIA contemporaries referred to as the Roadrunners for whom I am honored to have served as the alumni president for many years. These were the CIA, air force, and contractors working at or directly affiliated with the CIA projects at Area 51 named in this book. We share many memories of events at Area 51, and more when we gathered with our families on our boats anchored in our private, secluded cove on Lake Mead.

Lastly, I wish to acknowledge all my unnamed contemporaries, known and unknown who supported the CIA at Area 51 beneath the shroud of compartmentalized secrecy on projects more highly classified than the Manhattan Project that developed the atomic bomb.

Declassification

The Area 51 Chronicles is about the CIA's era at Area 51. It contains recently declassified material from the CIA's Top-secret/Sensitive Compartmented Information (TS/SCI) Kleyla report titled "The CIA, Directorate of Science and Technology [DST], History of the Office of Special Activities [OSA] From Inception to 1969." The CIA released this declassified report on 1 March 2016.

During September 1997, the National Air Intelligence Center declassified the formerly top-secret HAVE DOUGHNUT, HAVE DRILL, and HAVE FERRY FTD reported that the author references in this book.

On 16 August 2013, the CIA acknowledged its role at Area 51 with the release of "The Area 51 File" titled by the National Security Archive as, "The CIA Declassifies Area 51." The National Archives credited the author in the CIA document declassification of August 16, 2013.

https://nsarchive. wordpress. com/2013/08/16/the-cia-declassifies-area-51/

Preface

In 1955, it was not NASA, the US Navy, Air Force – it was the Central Intelligence Agency, CIA, who broke the ceiling in the high-flying evolution to put the U-2 manned reconnaissance plane soaring at 70,000 feet to escape the missiles of America's Cold War enemies. It was the CIA who advanced photography to gain high resolution at that altitude.

In 1957, it was the Central Intelligence Agency who pioneered stealth planes in the United States. In the 1960s, the agency produced the first operational stealth aircraft in the world.

In the 1960s, it was the Central Intelligence Agency who joined the Mach-3, 90,000-foot manned flight evolution to develop and fly the world's fastest and highest-flying manned air-breathing aircraft ever.

The Angels, the first book in the CIA Area 51 Chronicles series is my insider's account of the recently declassified legacy of how the CIA became the world's leader in secret military aviation technological and aeronautical engineering. Because it is about military planes, logically it should be about the United States Air Force, Navy, Marine Corps, Coast Guard, or National Guard. However, while this book is about these military services, it is primarily about the CIA being the genesis of Area 51, secret high-flying spy planes, stealth, and the exploitation of enemy assets. Recent declassifications by the CIA authorizes the author, an Area 51 veteran, to answer the questions Who, What, Where, When, and Why about the CIA activities in Area 51.

Book one of the CIA Area 51 Chronicles tells the genesis of the CIA and why it became involved in aerial reconnaissance. It explains how the United States entered World War II because of the lack of having an intelligence agency to foresee the Japanese attack on Pearl Harbor. It covers how and why President Truman formed the OSS, Office of Strategic Services with "Wild Bill" Donovan at the helm, and how the president disbanded the OSS less than a month after the surrender of the Japanese to end World War. The president and the military services feared a peacetime intelligence gathering agency.

William J. "Wild Bill" Donovan, the father of American intelligence and director of the Office of Strategic Services (OSS), the precursor to the Central Intelligence Agency, was an American soldier, lawyer, intelligence officer and diplomat during World War II. Wikipedia.

Introduction

Edwin Land

The Cold War began in the aftermath of World War II with the Soviet Union spreading communism and occupying countries that it liberated. That included Korea with Russia threatening to unify the two Koreas even before the American President Harry Truman removed all troops. The United States, emboldened the Soviet Union when the US removed Korea from its Pacific line of defense.

Making matters even worse, the US president also rejected Vietnam's plea for help in keeping colonialist France from returning to the now liberated Vietnam to resume colonialism. Instead, the United States withdrew the Office of Strategic Services, the OSS from Vietnam.

President Truman feared Wild Bill Donovan and his centralized intelligence, the OSS, and disbanded the service within a month of the end of World War II.

Consequently, the United States blindly entered the Vietnam War in 1945 by supporting France, paying 80% of the costs of France's war against the Vietnamese.

President Truman feared the spread of communism more than he did centralized intelligence, so in 1947, he formed the CIA under the National Security Agency and accountable to him, the president. The US Army, Air Force, and Navy treated the CIA as a centralized agency enacted by the President to collect intelligence gathered by the various military services. The military services feared the CIA and fought it at every opportunity.

Shortly after President Truman ordered all American troops withdrawn from the south of the 38th Parallel in Korea, the Soviet-backed North Korean army invaded South Korea. Thus, begin the so-called Korean War that Truman called a police action, not to admit his encouraging a proxy war between the United States and the Soviet Union.

Meanwhile, in Europe, the Soviet Union was sealing its borders to deny any intelligence gathering as it spread communism throughout Europe. The Soviet Union now had the atomic bomb, and the United States feared a nuclear attack. Additionally, the Soviet Union has placed the first satellite in orbit around the Earth and shortly afterward launches the first human into space. The United States saw the communist Soviet Union leading the arms and space races. It was imperative that the United States devise a means of knowing the extent of the Soviet Union advancements and its intentions. The threat of worldwide communism continued to grow.

In late 1951, the air force sought help from scientists to recommend the Strategic Air Command who needed a new way of conducting reconnaissance against the Soviet Bloc. Maj Gen Gordon P. Saville, the air force deputy chief of staff, added 15 experts in aerodynamics, propulsion, optics, and a broad spectrum of fields to an existing project on air defense known as project LINCOLN with the adopted code name, "The Beacon Hill Study Group." Comprising the group was the following.

James Baker from Harvard,
Edward Purcell from Harvard;
Saville Davis from the Christian Science Monitor,
Allen Donovan from the Cornell Aeronautical Laboratory,
Peter Goldmark from Columbia Broadcasting System Laboratories,
Edwin Land, Founder of the Polaroid Corporation,
Stewart Miller of Bell Laboratories,
Richard Perkin of the Perkin-Elmer Company,
Louis Ridenour of Ridenour Associates, Inc. and

Lt Col Richard Leghorn as the Wright Air Development Command liaison officer.

The Beacon Hill Study Group and others in Project GUSTO selected the Lockheed U-2 plane to fly reconnaissance flights over Russia under the management of the Dick Bissell of the CIA with the support of the air force at Area 51 in Nevada.

Development began on the U-2 in 1954 under the direction of a group headed by Richard M. Bissell of CIA. In June 1956, the plane became operational with a predicted useful lifetime over the USSR of two years.

Book One answered many of the questions, Who, What, Where, When, and Why. The CIA published the answers in the March 2016 declassification released as a report titled: CIA, Directorate of Science and Technology (DST), History of the Office of Special Activities (OSA) From Inception to 1969.

Neither Dick Bissell with the CIA nor Kelly Johnson expected the U-2 reconnaissance plane to be perpetually invulnerable to Soviet counter-countermeasures. No one expected the Soviet Union to be able to detect and track using the early American-built (Signal Corps Radar) SRC-584 radar system that Russia obtained through the World War II Lend-Lease program.

The research began to improve its survivability and extend the program's lifetime shortly after committing-the U-2 operationally in June 1956. The early studies grew to become a subproject of AQUATONE called Project RAINBOW. The CIA based its initial estimates of a high probability of success in U-2 overflights on the U-2's operating altitude. Its high penetration and operating altitude would diminish the possibility of hostile defense systems detecting and having an actual track of the U-2. Unfortunately, the Soviet air defense warning system not only discovered the U-2 by radar as it penetrated denied territory but quite accurately followed it over satellite and Soviet areas even during its earliest flights. The Soviet Union intensified its defensive efforts, which shortened the U-2's usefulness as a reconnaissance aircraft.

Thus, in July 1956, attention turned to the anti-radar research of Dr. Edward M. Purcell of Harvard University, who had discovered a possible means of countering or absorbing radar emanations. His discovery, led to laboratory work in techniques to blanket portions of the A-12 with radar absorptive materials to reduce radar detection. This technology could have significantly enhanced the U-2s prospects of continuing its reconnaissance role beyond the predicted eighteen months to two years.

Project RAINBOW laboratory's research and testing occurred under the auspices of a CIA proprietary research organization named as the Scientific Engineering Institute (SEI), Cambridge, Massachusetts. With the flight testing results by the firm of Edgerton, Germeshausen and Grier, Inc. (EG&G) at the Indian Springs Air Force Base, Nevada proving more promising than initially expected, the CIA deployed several RAINBOW-configured aircraft to Detachment B at Adana, Turkey in 1957. There, they had some degree of success in disrupting Soviet tracking of the U-2 on its missions. However, by mid-1957, obvious performance limitations became apparent as to radar camouflage of a conventionally designed and structured aircraft because of the weight and bulk of the absorptive material. Merely the weight of paint cost the plane 1,500 feet. The narrowband limitations of the camouflage technique could not cope with the frequency spread employed by the Soviet air defense warning system. Laboratory testing and measurement control system continued, but operational employment ended in August 1957.

A new approach became necessary after the failure to find a satisfactory solution to the radar problem using conventionally designed aircraft. Focus turned to the feasibility of a reconnaissance plane designed to a significantly reduced radar cross-section specification as the primary objective. Exploratory worked in this direction, and later efforts became known within the CIA as Project GUSTO.

NEVADA - THE BATTLE BORN STATE

 Pyramid Lake Torpedo Dropping Range

Wendover Air Base

 Stead Afb

Naval Air Base

Lovelock Gunnery

Reno Army

Nasa X-15 High Range

Naval Undersea Warfare Center

Navy Undersea Warfare Center

Nasa Nuclear Rocket Station

Marine Corp Mt. Warfare

Army Depot (World's largest)

Area 51

Nasa X-15 High Range

Atomic Energy Commission

  Indian Springs

 U.S. Coastguard

Indian Spring Afb

Nellis Afb

Chapter 1 - Replacing the U-2

In early 1955, Richard M. "Dick" Bissell, Jr., then SAPC (Special Assistant for Policy Coordination), Office of the DCI (Director, Central Intelligence) (SAPC/DCI) grafted a small project group into the staff of the Office of Special Activities, OSA to begin its organizational life. The reason was for the covert development and operation of the U-2 plane in conjunction with air force support. The CIA named the project with the cryptonym AQUATONE. Mr. Dulles and General Twining signed a formal agreement with the air force in June of that year, delineating areas of responsibility for both parties to the pact.

In June 1956, when the U-2 became operational, most officials predicted that the useful lifetime over the USSR no more than 18 months to two years. Its first flight over Soviet territory revealed the defense warning system not only detecting but also tracking it quite accurately. It stayed a unique and invaluable source of intelligence information for four years. All American manned flights over the Soviet Union ceased on 1 May 1960, when the Russians shot Francis Gary Powers down near Sverdlovsk.

Meanwhile, even as the U-2 started its active career, efforts underway made it less vulnerable. The hope reduced the vehicle's radar cross-section so that it became less susceptible to detection. Lockheed tried developments in radar-absorbing materials and achieved considerable success, though not enough to solve the problem. The CIA explored various far-out designs, most of them seeking to create a plane capable of flying at extremely high-altitudes. None of them proved practical.

Mr. Bissell, in a meeting with the Deputy Secretary of Defense, Mr. Donald A. Quarles, quoted from his prepared position paper to propose the course of action for Project GUSTO. His proposed program called for carrying forward studies, measurement, and experimentation with the RAINBOW camouflage or look at designing a new aircraft within three months' time. The scientific staff in Cambridge directed the technical work with actual systems, while AQUATONE project headquarters in Washington, DC kept the responsibility.

During this phase, specially selected manufacturers made contact as proper to explore the possibilities of unique materials and structures. They received the benefit of their views on the general design problem but kept continuous and intimate contact with relevant components in the air force and navy.

The CIA took steps to control discussions with manufacturers in the aviation and electronics industries. The CIA's concern was the issuance of formal requirements might stimulate unusual interest in conceptional non-radar reflective aircraft.

The priority was quickly selecting the best design approach for a low reflectivity reconnaissance plane. The job at hand meant evaluation with reasonable reliability, both its feasibility and its performance. It needed a governmental decision as to the advisability of a crash program to produce eight to twelve such vehicles.

Mr. Quarles, in a memorandum to the DGI on 26 November 1957, wrote that the Defense Department agreed with the purpose of the activities in Cambridge. He expressed a desire to take part in a definite design project decision at the proper time with a joint Agency DOD sponsorship that characterized policy and decision-making leading to the development of a follow-on reconnaissance platform.

In the fall of 1957, Bissell arranged an analysis of operations to decide the probability of shooting down an airplane at varying speeds, altitude, and radar cross-section. This study showed that supersonic speed significantly reduced the chances of detection by radar results.

The evidence showed that this line of supporting the supersonic line of approach might not reduce to zero the probability of the enemy shooting down our planes; however, it certainly needed serious consideration.

Therefore, Bissell increasingly focused his attention on building a vehicle capable of flying at extremely high speeds and great altitudes. Bissell still saw, from his efforts to hide the U-2 from radar, the need to incorporate, in the design, the best in radar-absorbing capabilities.

The CIA informed the Lockheed Aircraft Corporation and Convair Division of General Dynamics of the general requirements. Their designers set to work on the problem even before receiving any contract or funds from the government. From the fall of 1957 to late 1958 these developers continuously refined and adapted their respective schemes.

Bissell realized the exceeding expense of such development and production of such an aircraft. In the early stages, at least, it looked doubtful whether the project could succeed.

Before funding such a program, high officials demanded the best and most formal presentation of whatever prospects might unfold. So, Bissell selected a panel that included two distinguished authorities on aerodynamics and one physicist. He asked E. M. Land of the Polaroid Corporation to chair the panel.

Between 1957 and 1959 this group met six times, usually in Land's office in Cambridge. Lockheed and Convair sent their designers to some of the sessions along with the assistant secretaries of the air force and navy. All became concerned with research and development, together with one or two of their technical advisors. One beneficial consequence of the participation of service representatives was it reducing bureaucratic and jurisdictional feuds to nil. Both the air force and navy helped and cooperated throughout the process.

Allen Welsh Dulles, director of Central Intelligence from February 26, 1953, to November 29, 1961, under Presidents Dwight Eisenhower and John F. Kennedy. *Wikipedia.*

In early 1958, feasibility studies got underway on an advanced manned reconnaissance vehicle under the recommendations of the president's Scientific Adviser, Dr. James R. Killian. The President approved the recommendation of emphasizing the security requirements such an undertaking would impose. Director Dulles asked Mr. Bissell, who had spearheaded the CIA's U-2 Project AQUATONE and produced Area 51 in Nevada to take charge of replacing the U-2.

In May 1958, Mr. Bissell did so by forming an advisory panel chaired by Dr. Edwin K. with Polaroid and including the air force and navy's assistant secretaries for Research and Development.

The panel conducted a series of meetings in 1958 to consider the technical features need for an adequate successor to the U-2. The panel reported to Dr. Killian on its examinations with the mandate of making recommendations as to the type design meeting the requirements for the next generation reconnaissance plane.

As noted earlier, the technical direction of the new thrust in combating the Soviet electronic threat stayed at the Cambridge facility. The principals in the redirected effort were SEI, EG&G and Lockheed Aircraft Corporation (LAC). SEI turned its energies to theoretical aerodynamic models having the least radar cross-section characteristics. It engaged in wide-ranging experiments in shape control, type design, analysis, and alternative materials. Laboratory calibration and measurements conducted with radical and exotic model designs assessed their effectiveness in radar cross-section reduction.

As the months went by, the general outlines of what to do took shape in the minds of those concerned. Late in November 1958, the members of the panel held a crucial meeting. They now agreed it workable to build a plane where speed and altitude made it tough to track by radar. They recommended asking the president to approve in principle a further prosecution of the project and to make funds available for further studies and tests.

EG&G continued using the radar testing range at the Indian Springs Air Force Base, Nevada, where project headquarters had earlier, in support of project RAINBOW, installed a hydraulic lift, radars, antenna, and associated equipment. The company made its measurements on scale models raised on the hydraulic lift. Lockheed continued with preliminary design work on aircraft shapes and configurations during wind tunnel testing. The company tested the effect of materials and shapes for reflective characteristics and investigated substitute, non-metallic structures for use on portions of the airframe.

Lockheed subcontracted to Narmco, Inc., San Diego, California, for studies of the feasibility of certain types of plastic and high modulus fiberglass materials for use in the construction of the GUSTO vehicle. Also, Lockheed was going ahead with an independent configuration design study for a new aerial reconnaissance plane as a replacement for the U-2.

In Fort Worth, Texas, the Convair Division, General Dynamics Corporation proposed to employ a mother craft using the B-58 with the capacity of launching a small, manned reconnaissance plane.

There emerged in mid-1958 two general proposals, one from Convair and one from Lockheed. The former consisted of a high Mach, high-altitude, small, manned, ramjet-powered vehicle that launched from a B-58 bomber. Lockheed proposed a larger manned vehicle with high Mach, high-altitude, turbojet power initiated by the pilot. Both could achieve the desired operational specifications and within the development timeframe.

The Land Advisory Panel met on 31 July 1958 for the first time in Cambridge to obtain preliminary views on successor vehicles. The group received a briefing on the approaches undertaken by project headquarters, as well as on other advanced proposals from the air force. After reviewing all military aerial reconnaissance projects in being or in the study, the panel found it too early for judgment on the merits of the various ideas. No firm recommendations resulted from the first meeting. The board did not schedule another session for September 1958.

The committee did not make a design approach decision at the second meeting of the panel, feeling the need for a still further investigation. The group, at its September 1958 meeting, dropped any further consideration of project CHAMPION, which was a joint Agency, Navy feasibility study (paralleling GUSTO) of a possible high-performance reconnaissance plane.

Project GUSTO.

The president and his scientific advisor, Dr. James Killian already knew the situation. The CIA officials received a favorable hearing when they went to them with the recommendations of the panel. The president gave his approval and asked Lockheed and Convair to give formal proposals. He made funds available to them, and the project took on the codename GUSTO.

They completed the two proposals less than a year later and on 20 July 1959; they briefed the president again. This time he gave final approval, which signified the program was getting fully underway.

The next significant step chose between the Lockheed and Convair designs. On 20 August 1959, they gave specifications of the two proposals to a joint DOD/Air Force/CIA selection panel:

The committee selected the Lockheed design, which concluded Project GUSTO and began Project OXCART, the program to develop a new U-2 follow-on aircraft. On 3 September 1959, CIA authorized Lockheed to continue with antiradar studies, aerodynamic, structural tests, and engineering designs. On 30 January 1960, the CIA approved producing 12 planes.

The proposed inflated vehicle, ramjet-powered to Mach 3, reaching 125,000 feet altitude was a radical departure from the conventional aircraft design. Nonetheless, a NACA study strongly recommended that the US Navy pursue it. Convair, Boeing, Hughes, Marquardt, and Goodyear Aircraft Corporations had all conducted studies about CHAMPION, and while the proposal appeared workable, the panel did not like the five years estimated to develop the system.

The panel held a final meeting in Boston on 12 November 1958, and reported its findings to Dr. Killian on 15 November as follows:

a. The successor reconnaissance plane had to achieve a substantial increase in altitude and speed be of reduced radar detectability suffer no loss in range to that of the U-2 and be of the smallest size and weight,

b. The panel concluded that the most satisfactory design approach was the small, lightweight aircraft launched from the B 53. The possible problems foreseen were aerodynamic heating and in the air inlet system. As a second and less desirable choice, the panel selected a similarly small, lightweight plane capable of unassisted takeoff, but with slightly less speed and less than desired range.

c. The panel recommended system development on an expedited and secure basis with the prerogative of reviewing alternative systems should the panel's choice prove unacceptable

The panel received Presidential approval for Project GUSTO investigations and centered its interest on Lockheed's proposal for a supersonic, high-altitude unstaged design and Convair's design proposal for a parasite to the B-58 aircraft. The Convair design configuration especially minimized radar return, whereas the Lockheed design made no concessions in this, which compromised the aerodynamic performance.

Convair began work in December 1958, on a contract which called for original studies, tests, and preliminary design of a high-altitude, supersonic reconnaissance vehicle to replace the U-2. It was to be a four-and-one-half month, engineering effort and funded by $1,200,000. A $1,000,000 contract let with Lockheed concurrently for similar studies.

The Marquardt Aircraft Company, Van Nuys, California began preliminary, engineering design studies and tests necessary to develop a ramjet engine compatible with the airframe design proposed by Convair. All of this became a four-and-one-half-month endeavor with studies and tests resulting in model specifications of the engine and engine controls at an estimated cost of 2,500,000. To this point, Lockheed's designing had considered only non-existing turbojet propulsion systems, not under development by the air force or navy.

The panel solicited additional studies from manufacturers of camera equipment, electronic equipment, arid pilot protective assembly systems. It focused on such things as pressure suits, and oxygen sources. EG&G extended its work at the range to provide a testing capability for the cross-section models furnished by the two airframe contractors. The panel continued consultant services with Narmco and SEI during a phase of Project GUSTO that terminated at the end of June 1959, at a total estimated cost of $5,420,000.

Comparison of the Major Design, Features

	Lockheed	Convair
Aircraft designation	A 3	FISH
Speed	Mach 3.2	Mach 4. 2
Range	3,200 nm.	3,900 nm
Altitude	90,000 ft.	90,000 ft.
Launch system	Pilot launch	From B-58
Propulsion system.	2 turbojets	2 ramjets
Weight	95,000 lbs.	38,500 lbs.
Predicted first flight	January 1961	January 1961

The lack of knowledge and experience in the ground handling equipment, fuels, and retrieval procedures, requirements, made it difficult from a coordination view to compare the two vehicles. The A-3 was easier to handle on the ground. However, the Convair design showed superior performance criteria.

During the spring of 1959, Lockheed and Pratt & Whitney (P&W), were one team, and Convair and Marquardt were the other as they continued their design, model construction and testing, structural investigations, and another testing. The progress of both systems closely monitored and reviewed by a joint CIA - air force evaluation team. Both teams revised plans and designs frequently, until May 1959 to produce a summary comparison of the two Systems.

The Convair FISH design called for a relatively small vehicle with a gross weight of roughly 40,000 pounds. The FISH design called for staging its entry from a B-58 mother aircraft and having a 4,000-mile range at a 90,000-foot altitude. Two 40-inch diameter Marquardt ramjets powered the FISH using JP type fuel. The roughly 50 feet length and 35 feet wingspan design minimized radar reflectivity by configuration and utilization of preferred materials. The FISH had two small turbojets incorporated into the subsonic portion of the flight to follow the ramjet powered supersonic flight segment. The current B 58A model was unable to exceed the transonic to reach the Mach speed required for efficient ramjet operation (2.7 Mach) was its most severe deficiency. The next model, the B-58B, would have sufficient power to accomplish the task. However, at this point, the inlet and engine testing had not proceeded far enough to surface major problems.

The Boston meeting did not result in a decision regarding the two vehicles. However, the participants did agree that from an operational viewpoint, the desirability of the A-11 with its 4100-mile range and increased altitude capability. It also had the advantage of a conventional take off and the ability to use a short runway. However, its principal design deficiency allowed radar track.

The Convair FISH being a staged vehicle made it a much more complicated system to operate. However, the possibility of its being able to fly missions undetected was greater than that of the A-11. The aircraft design and radar cross-section and not operational problems chiefly concerned the technical experts on the Land Panel. The meeting ended with the conclusion to expect sporadic detection and tracking by radar regardless of vehicle. The panel made no recommendations as to the choice of aircraft before the CIA/air force briefing team returned to Washington to gain presidential approval to continue the GUSTO program.

Systems selection meetings quickened the tempo of GUSTO activity during June 1959 with the contractors submitting design changes and new approach concepts. The air force gave Convair a severe setback when it canceled procurement of the B 58B. The small Convair design staged from the B 58. Radar testing showed good results and aerodynamic testing and structural development proved successful.

The older B 58A model mother aircraft now ruled out the FISH design because of the cost and operational complexity of reconfiguring it with two additional engines necessary to achieve the speeds required for efficient ramjet engine ignition on the FISH vehicle. Additionally, the air force was extremely reluctant to reduce its small inventory of advanced 58As bombers for modification to a drone mothership.

General Cabell, Mr. Bissell, Purcell, Land, Drs. Killian, George B. Kistiakowsky, Bruce Billings, and Franklin A. Rodgers received a thorough review on Project GUSTO at a meeting in Mr. Dulles' office on 14 July 1959. They decided that neither the Convair FISH design nor the Lockheed A-11 design met their criteria for replacing the U-2.

Both Convair and Lockheed were interested in submitting new design proposals with designs powered by Navy P&W J58 turbojet engines and offering reduced radar return characteristics.

Everyone accepted that a compromise was necessary between radar reduction attempts and maintaining a good aerodynamic design. The group recommended continuing GUSTO. However, they wanted the two manufacturers to submit new design proposals. If they got the concurrence of the secretaries of defense and the air force, the secretaries and the air force would communicate their joint view to the president.

General Cabell and Mr. Bissell first reported on GUSTO's status to Air Force Secretary Douglas and General White. On 15 July 1959, they briefed the Secretary of Defense McElroy and Under Secretary Gates of same and received a unanimous recommendation to urge the president for the continuation of the program.

The president received a briefing on 20 July 1959 by Mr. Dulles, General Cabell, Mr. Bissell, Land, General White, Secretary McElroy, Drs. Killian, and Kistiakowsky. The president approved the study's approach toward gaining intelligence on the Soviet Union and instructed Mr. Bissell to work with the Bureau of the Budget on the funding essential to the continuation of the effort. The choice of contractors hinged on the final design proposal submissions.

A meeting between Mr. Bissell and the Bureau of the Budget personnel on 22 July 1959 ended with the understanding that necessary financial arrangements were forthcoming to carry on the program.

Now the one major step of which design proposal to pursue remained before entering a full-scale development agenda. Lockheed and Convair had both submitted new proposals mid-August 1959. Both were unstaged aircraft differing only in external configuration. Both proposed aircraft would reach an altitude of 90,000 feet, fly at Mach 3.2 with an approximate 4,000-mile range. Both had a similar size, weight, and aerodynamic performance and preferred the P&W J58 engine over the General Electric Corporation J 93 which lacked higher cruise altitude of the J58.

17 August 1959 Comparison of general, characteristics.

	Lockheed	Convair
Aircraft designation	A-l2	KINGFISH
Speed	Mach 3.2	Mach 3.2
Range (total)	4120 nm	4000 nm
Range (at altitude)	3800 nm	3400 nm
Cruise Altitudes	Lockheed	Convair
Start	84,500 ft.	85,000 ft.
Mid-range	91,000 ft.	88,000 ft.
End	97,600 ft.	94,000 ft.
Dimensions		
Length	102 ft.	79. 5 ft.
Span	.57 ft.	56. 0 ft.
Gross Weight	110,000 lbs.	101, 700 lbs.
Fuel Weight	64, 600 lbs.	62,000 lbs.

Lockheed's designer, Clarence L. (Kelly) Johnson, creator of the U-2 called his new vehicle the A-11. Its design exhibited many innovations.

The designation changed to A-12 to distinguish it from the A-11 designator, for the all-metal version proposed initially. Small-scale testing predicted the Convair being slightly better at S-Band frequencies.

On 20 August 1959, the joint DOD/air force/CIA source selection group chose the Lockheed design. Initial development, exclusive of engine costs, went to Mr. Bissell at project headquarters with the continuation of the Lockheed arrangement beyond initial development. Continuing the project was contingent on the success of design changes in the A-12 reducing the radar cross-section.

The two factors favoring the choice of Lockheed were its substantially lower bid and the company's experience from the U-2 program. Lockheed was already geared to launch into another highly classified program. It had handled the U-2 program without attracting undue attention in the industry and still possessed a reservoir of labor with the necessary security clearances and was readily available. Lastly, everyone was confident in Mr. Clarence L. (Kelly) Johnson and his ability to produce a new vehicle as he had done with the U-2.

All persons associated with GUSTO received notification terminating all activities performed under that title effective 31 August 1959, thus minimizing the possibility of speculation about the creation of a follow-on program. On the second floor of CIA headquarters, Bissell set up an OXCART department restrictive for anyone lacking a need-to-know of a project seeking to replace the U-2 that was still more highly classified than was the Manhattan Project that developed the atomic bomb.

Developing the OXCART

Supersonic airplanes, however, involved a multitude of tough design problems. Their payload-range performance is highly sensitive to engine weight, structural weight, fuel consumption, and aerodynamic efficiency. Small mistakes in predicting these values often led to significant errors in performance. Lockheed tested and retested models of the A-11, adjusting and readjusting, during thousands of hours in the wind tunnel. Johnson felt confident of his design. However, no one could say positively whether the bird flew, much less fulfill the extremely demanding requirements laid down for it.

To make the drawings and test the model was one thing; to build the aircraft was another. The most numerous problems arose from heat. The design speed of the plane subjected it to a temperature of more than 550 degrees Fahrenheit. No metal hitherto commonly used in aircraft production stood this temperature. The weight of those metals can stand the heat made them unsuitable for the purpose in hand.

The CIA immediately acted on the decision to proceed on a limited basis with the Lockheed design proposal, establishing a new operational security clearance category like the U-2 program. The new project called "OXCART" was born with a demonstrated "need to know" requirement for obtaining an OXCART clearance. Project director, Mr. Bissell retained the authority to approve clearances for military and civilian personnel. The project Security Staff handled the security clearance of the industrial and supplier personnel by coordination with the technical management group. In the early days of the program, the CIA cleared and briefed only the top ranking and critical Agency, DOD personnel and elsewhere in government who could make a direct and needed contribution.

On 3 September 1959, the CIA issued a letter contract to the Lockheed Aircraft Corporation to authorize proceeding with aerodynamic, antiradar studies, structural tests, and engineering designs for the proposed A-12 aircraft. The CIA obligated four and one-half million dollars against FY 1960 funds. The contract called for the construction of a one-eighth scale model for radar cross-section (anti-radar) (AR) testing at the Indian Springs facility. The agreement called for the development of required testing facilities, a full-scale model for follow-on anti-radar testing. Other requirements in the agreement included the construction of an aircraft section of titanium, wind tunnel testing, and so forth. The program included the production of twelve aircraft the contractor, reducing radar reflectivity, providing construction methods, and affirming that the design met the desired specifications for an advanced reconnaissance system. Production of aircraft was subject to negotiation later.

During the design phase, Lockheed evaluated many materials and finally chose an alloy of titanium. The titanium provided great strength, relatively lightweight, and excellent resistance to high temperatures. However, they found titanium scarce and very costly. With no fully developed methods of milling it and controlling the quality of the product. Lockheed rejected some 80 percent of the early deliveries from the Titanium Metals Corporation.

In 1961, a delegation from headquarters visited the company officials to inform them of the objectives and high priority of the OXCART program to gain their full cooperation. The supply became consistently satisfactory.

The visit solved an initial problem. One of the virtues of titanium was its exceeding hardness. However, this very attribute caused immense difficulties in machining and shaping the material. Lockheed devised new drill bits to replace standard bits that worked well on aluminum but soon broke to pieces when used on titanium.

Problems such as those made assembly line production impossible; requiring Lockheed to build the small OXCART fleet by hand. The cost of the program mounted well above original estimates and soon ran behind schedule.

One after another, however, the OXCART participants solved the problems, and their solution constituted the greatest single technological achievement of the entire enterprise. Henceforth it became practicable if expensive, to build aircraft out of titanium.

The plane's critical weight eliminated insulation to solve the problem. The inside of the plane heated like a moderately hot oven, requiring the pilot to wear a kind of space suit, with its cooling apparatus, pressure control, oxygen supply, and other necessities for survival.

The fuel tanks, which constituted by far the greater part of the aircraft, heated up to about 350 degrees. The plane required a special fuel and rendering the tanks inert with nitrogen lubricating oil formulated for operation withstood 600 degrees F. and contained a diluent to remain fluid at operation below 40 degrees. Insulation on the plane's complex wiring soon became brittle and useless. During the lifetime of the OXCART, no one came up with better insulation. The wiring and related connectors got special attention and handling at great cost in labor and time.

The CIA Reopens Watertown at Area 51

Naturally, the CIA needed a secret location for testing Lockheed's triple-sonic A-12. The CIA considered ten air force bases programmed for closure. None provided adequate security and presented prohibitive annual operating costs. The CIA and Lockheed selected Groom Lake even though it lacked personnel accommodations, fuel storage, and a suitable runway.

Dramatic changes came to Area 51 two years later with the appearance of CIA Project OXCART to develop the Lockheed A-12 proposed a successor to the U-2.

The OXCART A-12 aircraft was a sleek, sturdy looking aircraft with a long tapered forward fuselage with blended chines. A rounded delta wing supported two turbo-ramjet engines capable of boosting the aircraft to Mach 3.2 at altitudes more than 90,000 feet.

Twin inwardly canted tails and an internal sawtooth structure in the wing edges contributed to a low overall RCS. Its titanium airframe contained asbestos-fiberglass and phenylsilane composites in the leading and trailed edges, chines, and tails for RCS reduction. The number 12 represented the final, and the "A" stood for "Archangel."

The A-12 Mach 3, high-flying reconnaissance plane as America's first plane designed for stealth required a radar cross-section (RCS) test capability, a need giving birth to a special projects exploitation team identified simply as special projects. Edgerton, Germeshausen & Grier (EG&G), the prime contractor for the AEC atomic bomb testing established the organization for the CIA at Groom Lake.

Base Facilities

Note: People, the CIA and the author included, commonly refer to the CIA's flight testing facility at Area 51 as a base. However, this is not correct. It was a facility with no official name. The facility became a base when the air force took over the facility in 1979.

The OXCART aircraft program based at Area 51, a restricted area on the Nevada Test Site, had the necessary facilities and staffing to support the test, training operations and operational deployment of the A-12. The population would soon average 1,500 persons, including military and CIA civilian employees, on station to support the OXCART and TAGBOARD projects. About 650 of these were in direct support of launching operations, and approximately 611 were involved in indirect support such as coordination, firefighting, and guards. Most of these people were under contract to the Lockheed Aircraft Company or its subcontractors and were on permanent duty in this area. The military personnel and CIA civilian employees are on a basic three-year tour.

The CIA invested a total of $21 million in Area 51 for runways, buildings, housing, navigational aids, and water supply, to make the base self-sufficient. CIA personnel supervised the base support and maintenance. Reynolds Engineering and Electrical Company, a contracting company from Las Vegas, had 239 persons engaged in base maintenance work. Total cost per year for salaries and the necessary equipment was $5.5M.

The required conducting of radar tests on a full-scale model raised the issue of the Indian Springs hydraulic lift's inability to raise and lower the full-scale model. Nor did the Indian Springs facility afford the necessary security to prevent the model being seen by unauthorized persons. The CIA decided, therefore, to move EG&G's anti-radar testing equipment from Indian Springs to Area 51 and to install a heavier hydraulic lift and pole device to accommodate a full-scale model and, eventually the aircraft itself. The CIA chose to return to the former U-2 Watertown facility in caretaker status for reasons of security, access, and accommodations,

EG&G continued scale model testing at Indian Springs while the CIA prepared to reopen Watertown. By October 1959, the Atomic Energy Commission had reactivated, and in late November EG&G had its test range equipment moved from Indian Springs, and the Watertown site ready for anti-radar tests of the full-scale A-12 mock-up. The CIA placed a staff employee in command of a contractor population of 75 persons to begin operations at Area 51 on a crash basis and under austere conditions.

Meanwhile, the Development Projects Division (DPD) had moved forward to establish a system of security and to develop cover stories to explain the new activity at Lockheed (as well as at other contractors' plants), and the reactivation of Watertown. This venture was into beyond state of the art in aircraft development. If known, in the aerospace industry, it would quickly lead to speculation as to its real purpose and inevitable compromise of the concept.

Setting the Pecking Order

As with the earlier U-2 program at Area 51, a new memorandum of understanding was necessary with the air force to delineate areas of responsibility between the CIA and the air force. In a letter to the director, Central Intelligence on 18 September 1959, General White, Chief of Staff, Air Force, assured the CIA of the air force's continued interest in the project on the same joint basis as in the U-2 program.

In this regard, General White referred to the original agreement of August 1955 which outlined and defined in rather broad terms the areas of responsibility between the CIA and the air force. He agreed to the previously agreed intent of the basic concepts and organizational structure. He directed Col Geary to meet with the CIA's designated representatives to review the original document and suggest such changes or additions mutually acceptable and beneficial.

The CIA wrote a classified contract with the Pratt & Whitney Division of United Aircraft Corporation to provide the propulsion system for the aircraft.

The US Navy had originally sponsored the development of the J58 engine for its purposes. The Navy's interest in the J58 development, however, had subsided, giving the air force assumed sponsorship of a proposed advanced Air Force weapon system with the J58 engine extending to highly classified Mach 3.2 performance at extremely high-altitudes. The terms of the contract called for the assembly of three advanced experimental engines for durability and reliability testing, and possibly experimental flight testing in early 1961.

Pratt & Whitney

The Pratt and Whitney Division of United Aircraft Corporation participated in discussions of the project and undertook to develop the propulsion system. The Navy originally sponsored for its own Mach 3 uses the J58 engine proposed for the A-12. Navy interest in the development diminished and the secretary of defense decided to withdraw from the program at the end of 1959.

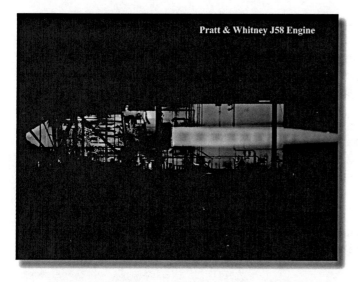
Pratt & Whitney J58 Engine

The CIA required the engine and aircraft be further developed and optimized for a speed of Mach 3.2. The new contract called for an initial assembly of three advanced experimental engines for durability and reliability testing, and provision of three engines for experimental flight testing in early 1961.

Two J58 P&W turbojet engines powered the Lockheed design, now designated the A-11. The A-11 design was 100 feet long, with a wingspan of 50 feet, and weighed 92,000 pounds at takeoff. Its design called for flying 4,100-miles at 85,000 to 95,000 feet altitude. However, it was more susceptible to radar detection because of its size and had a more severe sonic boom effect than the smaller Convair design. Thought Lockheed did not anticipate serious aerodynamic problems with the engine installation, it needed to uprate the J58 engine to a Mach 3.2 performance and to develop the necessary afterburner. Now, Lockheed was proceeding to fabricate and test structural components using titanium.

On 18 May 1959, the CIA and the air force met to discuss Project GUSTO's status and future planning. Attending were the Air Force Gen Jacob E. Smart, Assistant Vice Chief of Staff, General Thomas D. White, Chief of Staff, and Air Force Colonel Leo P. Geary for the air force. Mr. Dulles, General Cabell, Mr. Bissell, and Colonel William Burke attended for the CIA. After reviewing the status of both proposals, Mr. Bissell pointed out the advantages and disadvantages of each, noting an upcoming Advisory Panel meeting in Dr. Land's office in Boston in early June. At the June meeting, they expected to make recommendations regarding the proper course to take. At that time, the Department of Defense would review the panel's recommendations, and make a final system selection to take to the president for approval. At General White's request, they formed a joint air force/CIA working level technical panel to provide the skills in final system selection.

The CIA realized the long lead time involved in the manufacture of jet engine components when it contracted with P&W for delivery of hardware before deciding to proceed with the production of airframes. Meeting Lockheed's projected airframe delivery schedule required deciding to build engines at this early date should a decision be made to proceed with production of A-12 aircraft.

The J58 engine powered the A-12 with 32, 500 lbs. of thrust. It rated 100 hours (military time) between overhauls and had a growth potential to 150 hours between overhauls.

The Cameras

Then there was the unique problem of the camera window. The OXCART carried a delicate, highly sophisticated camera, which looked out through a quartz glass window. The effectiveness of the whole system depended on achieving complete freedom from optical distortion despite the high heat exposure to the window. The heat issue could not be solved just by providing equipment with resistance to heat. The temperature had to remain evenly distributed throughout the area of the window.

It took three years of time and two million dollars of money to arrive at a satisfactory solution. The program scored one of its most remarkable successes when the quartz glass successfully fused to its metal frame using an unprecedented process involving the use of high-frequency sound waves.

During the GUSTO program, camera manufacturers, Fairchild Camera, Hycon, Perkin Elmer (PE), and Itek had proposed engineering design that project headquarters had an evaluation team from and the CIA's Photographic Interpretation Center review. The CIA concluded in early 1959 that the Perkin Elmer model was the best, and recommended choosing it for the successor reconnaissance system. The Land Panel affirmed this view after reviewing proposals. Perkin Elmer was given a new contract in October 1959 to engineer and design for a photographic system for the OXCART A-12 program.

Thus, Perkin-Elmer became the primary camera manufacturer. However, the extreme design complexities forced a decision for a backup system in the event the Perkin-Elmer model ran into production problems. The CIA also asked Eastman Kodak to build a camera.

Firewel

The Firewel Corporation had developed the life support system for the U-2. Now, the CIA tasked the company with building some of the A-12 life support systems. These included the parachute system and its enclosed life support system. Firewel built the test chamber at their plant. The high-altitude simulation chamber was somewhat similar A-12 cockpit enclosure, redesigned to provide the low pressure/high-temperature conditions to test and ensure that the new suit "wearer" felt confident in the suits capability/safety.

The CIA declassification of the U-2 and A-12 programs revealed Firewel Corporation as another of the startup companies and obscure companies fronting for others in the interest of secrecy and national security. Firewel had pioneered the development of high-altitude oxygen systems, testing them on monkeys and advancing to pilots in the X-15, U-2, and A-12.

In 1946, two Scott Aviation alumni glanced around the basement of their home at 135 Aurora Street in Lancaster, New York and decided it was a good place to launch a company they named Firewel. Philip Edward (Ed) Meidenbauer, Jr., the president, Donald Nesbitt, a ceramics engineer, was the vice president. Ed, a self-taught mechanical engineer, had been Director of Oxygen Research at Scott Aviation, where he developed the original Air-Pak. Ed's brother, Clifford Meidenbauer, a Signals Corp officer during World War II, joined Ed and Donald as the financial officer of their new company, Firewel.

Operating from the basement amidst Lois Meidenbauer's home-canned peaches and jellies, near the old furnace, and around the corner from her laundry room, the company began building furnace burners that converted old coal furnaces to oil or gas.

Workers trooped into the house and down the cellar stairs where they hammered out the burners. Buyers and suppliers conducted their business with the company from the living room sofa, often charmed by Ed's youngest daughter who loved to toddle to them and crawl into their laps. By 1947 the company had advanced from coal furnace conversions to a full line of furnaces.

On the outside, occurring events eventually changed the company drastically. In March 1946, the first US-built rocket left earth's atmosphere, reaching an altitude of 50 miles at about the same time Ed was leaving Scott. A month later in April, the US Navy revealed it had created an 8,000-horsepower aircraft rocket engine. On August 27, the Navy successfully tested a pilot ejector seat at Wright Field.

In October 1947, the Air Force became independent of the US Army, and on the 14th air force test pilot, Chuck Yeager broke the sound barrier, reaching 600 mph in a Bell X-1 rocket plane. On February 28, 1948, Yeager exceeded the speed of sound again in a Bell XS-1. Moreover, on August 8, 1949, pilot Frank Everest climbed to 63,000 feet in the Bell X-1.

Because of their technical background, Firewel's principals decided to add a corporate focus on the future of high-altitude military flight for needed innovative kinds of breathing apparatus. Ed, Don, and Cliff saw this as a means of expanding their business and contacted various individuals they had known during the war. Their original development contract was for $80,000 followed by an additional $70,000 extension.

In 1951, Ed and Firewel contacted to solve problems with the regulators for the prototype Navy space suit developed by David Clark Company and the BF Goodrich Company.

Designing and manufacturing small valves, regulators, and system, Firewel advanced application and technology of an aircraft-mounted regulator to pressurize the face-piece and capstans of early partial pressure suits.

The partial pressure suit provided counter pressure to the torso and face-piece of a pilot if exposed to the barometric pressure at altitudes of 40,000 feet or higher. Firewel equipped the spacesuit with the instrumentation and controls, oxygen breathing and ventilating systems that automatically protected the wearer from the elements and hazards of space. David Clark Company designed the suit for such hazards as blood-draining acceleration, blood-boiling low pressure, and from cosmic rays and extreme temperatures.

Firewel designed the Model 1-A, a prototype stratospheric suit with a Firewel regulator that supplied an oxygen and air mixture on demand or pure oxygen under pressure through a mask attached to a standard naval aviator's crash helmet.

In 1952, seven years after laying much of the patented technology groundwork, President Ed Meidenbauer died. Donald Nesbitt succeeded him with Clifford Meidenbauer continuing to handle financial matters. By 1956, Firewel employed 140 people and was a multi-million-dollar corporation.

By the mid-1950s, Firewel's specialty of high-altitude regulators exceeded the production of all their larger competitors. The company's aeronautical division with only one-third of the employees accounted for almost 70 percent of the firm's business. In part, this attributed to Firewel's new concepts in oxygen equipment and survival system integral to the pilot's flight gear and part of the state-of-the-art military planes developed at the time.

Firewel had designed new valves and incorporated them into the pilot's parachute pack rather than the instrument console. Firewel used miniaturized oxygen hoses and radio communications wiring and developed delicate silicon diaphragms strengthened with nylon filament.

The switch from regular to bailout oxygen was automatic, a new concept in oxygen equipment and survival systems. Now, the equipment became an integral part of the pilot's flight gear.

Firewel first developed the miniaturized mask-mounted oxygen system for the US Navy pilots flying the A-3D Skywarrior, F-4D Skyray, and the F-8U Crusader.

Firewel became a major player with the air force and navy in the development of breathing apparatus for high-altitude flight.

The Firewel-built survival kit was a soft pack interfaced with the aircraft oxygen supply and connected to the pilot by two hoses. One hose provided breathing oxygen from the miniature regulator to the facepiece and chest bladder, the other attached to the capstans of the partial pressure suit. They designed the microphone and earphone wiring harness into the hoses of the survival kit as well as the power to heat the face-piece of the partial pressure suit. The emergency oxygen supply in the survival kit manually actuated by pulling a cable that terminated in a green ball commonly referred to as "the green apple."

They used the soft pack survival kit because the earlier versions of the U-2 aircraft did not have an ejection seat. The soft pack survival kit after that evolved into a rigid hard-shell kit used by all century series aircraft. Field service representatives trained and deployed from Area 51 "Watertown" with the original three detachments of the U-2 programs in 1956. The air force similarly staffed subsequent deployment of air force programs at all locations as these programs evolved.

By the late 1950s, the Firewel production line was producing a 100-man oxygen regulator, oxygen bottle and mask, anti-g gloves, and a "Global Survival Kit," for pilot ejection seats. They soon added pressure regulators for flight suits and helmets, a backpack oxygen manifold, "Get Me Down" oxygen supplies, a back-pan oxygen assembly, disconnect couplings, oxygen gauges, and a space vehicle pressure reducer.

It was during this period that the company became involved with the design of high-altitude breathing apparatus and life support for the ultra-secret Lockheed U-2 CIA spy plane test flown at Area 51.

After 12 years as a discrete company, Firewel with 440 local employees sold in 1958 to the Aro Equipment Company of Ohio.

In 1959, Aro-Firewel designed environmental regulating system successfully sustained monkeys Able and Baker in a space capsule launch to research biological effects of flight. Neil Armstrong in the X-15 tested the company's oxygen high-altitude regulators used since 1951 in every record flight.

In the early 1960s, ARO/Firewel again contracted to design and develop a complete respiratory life support system for the various versions of the Blackbird. Specifications for the oxygen supply systems and pilot respiratory equipment required redundancy.

Therefore, Firewel created complete dual systems that included dual oxygen supplies, a dual oxygen control panel, a dual ejection seat disconnect, dual supply hoses to the pressure suit, a full pressure suit controller with redundant backup, and a dual helmet mounted breathing regulator.

Additionally, the survival kit contained dual emergency oxygen supplies, regulators, and hoses that interfaced with the pilot. The pressure suit employed on these aircraft was of the full pressure type. It encapsulated the pilot in the pressure suit that maintained an absolute pressure of 3.5 psi at altitudes of 35,000 feet and higher.

Total encapsulation necessitated the need for body cooling. Therefore, Firewel provided an adjustable flow control valve to regulate the engine bleed air used for suit ventilation and incorporated dual emergency oxygen supplies into the rigid survival kit. Actuation automatically occurred upon ejection or by manual control. The aircraft oxygen supply initially specified as compressed, high pressure converted to low-pressure liquid oxygen ARO/Firewel also provided large 100-liter liquid nitrogen systems used to inert the fuel.

In 1968, Aro moved production to its Ohio headquarters, but let the aerospace engineering and development entity remain in western New York. In 1985, Todd Shipyards bought the business from Aro. Ingersoll Rand took over the group in 1990. Carlton Controls, founded by a former employee, purchased the aerospace aspect of the firm in 1993 and continued to produce devices for high-altitude and space flight.

Firewel subjected the A-12 pilots to a simulated six-hour mission in this chamber. Part of this test mission build-up was the installation of some personal instrumentation to allow monitoring blood pressure, EKG, and internal body temperature. This testing both proved the suit and gained the pilot's confidence in the equipment. An inserted anal probe measured the body temperature.

Some suit training occurred at Area 51 for the project pilots using the base swimming pool as the scene of some "dunk" training. This involved the pilots jumping off the diving board with a full survival kit attached to the pilot/suit to acquaint the pilot in the steps required when coming down the water.

The Firewel Corporation continued its studies in the pilot environment, as did Narmco with its work in plastics and other radar absorbent materials at the request of the CIA. Eastman Kodak agreed to a feasibility study for a camera designed to OXCART operational characteristics. EG&G was performing the anti-radar testing on the OXCART model, and SEl was providing consulting services about the radar cross-section reduction features of the A-12 system,

As the new year approached, developments in the OXCART program were reaching a stage when a final decision was required to engage, in production. At an OXCART Suppliers' Meeting on 16 December 1959, Mr. Johnson stated that Lockheed needed a configuration freeze and go ahead with an accurate production figure to plan and implement tooling, requirements efficiently. Mr. Bissell agreed to make this the principal subject, at the next such meeting in mid-January 1960 along with the radar cross-section problem.

On 20 January 1960, Mr. Bissell, with Agency and the air force representatives, met the Land Panel consultants to review the status of the radar cross-section of the A-12. The panel was impressed with the progress made in the radar cross-section areas. Dr. Frank Rodgers of SEI predicted continued testing, achieving further improvement in reducing radar returns. This was a condition necessary to proceed to the final design production decision.

However, at the OXCART suppliers progress review meeting the next day, Mr. Bissell focused on an increasing weight problem creeping into the A-12 design and the resultant altitude and range degradation. Mr. Johnson was directed to investigate the weight problem to provide a specified minimum range and altitude profile of the A-12. The CIA also asked him to give estimates of A-12 performance at Mach 3 5 and advised him that until the information was in hand and evaluated, there would not be a decision concerning the go ahead.

Several days later, Mr. Johnson reported the weight reduction and minimum mission profile was considered acceptable with slightly improved range arid an extra 3,000 feet of altitude from the Lockheed estimate of A-12 performance at a speed of Mach 3.5.

On 30 January 1960, the CIA gave Johnson a firm go-ahead to produce 12 aircraft, completing the three-year search for a manned vehicle successor to the U-2 reconnaissance system that began in August of 1957. To this point, the OXCART program had expenditures totaled $49,497,366 for FY I960 (44 million for engine development and initial production by P&W). Additional work and services resulting from the decision to build the A-12 reconnaissance plane were estimated to be an approximate 44 million more for a total FY I960 program approval of 93,780,000. Before releasing the funds, the Bureau of the Budget raised the question of additional approval by the president for the continuance of the program and was assured such was not the case.

At the 20 July 1959 meeting with the president, the president had said that he doubted the CIA able to attain some of its technical goals. However, he further stated if the CIA met these goals the project could proceed, Mr. Bissell had determined, from the last professional, reviews, that the CIA had complied with the president's guideline. He determined that it was not necessary for the CIA to return to the White House for further approval. On 8 March I960, the director signed such a memorandum to the Director of the Budget.

The CIA gave the prime contractor, Lockheed as much latitude as possible to ensure the company met the schedule when it forecasted a schedule for delivery of A-12s. The first airframe delivery was April 1961 and the last in August of 1962. The CIA authorized Lockheed to dispense with competitive bidding for airframe subsystems procurement. However, the company had to provide reasonable and prudent judgment in awarding such contracts. The CIA gave this authorization for reasons of security, timeliness in procurement, and overall system compatibility. However, Lockheed was required to maintain detailed records regarding selections and evaluations.

For this new program at Area 51, both the CIA and Lockheed relied on the philosophy of management developed and employed successfully in the U-2 program. The philosophy had been to select contractors carefully and to grant them maximum, technical responsibility and authority to get the job done by the contract terms and functional specifications. When faced with significant technical choices or changes the contractor would refer the matter to the project director for advice and decision on anything having an impact on funding, schedules, or performance.

Headquarters would monitor the technical progress of the contractors using periodic reporting, supplier conferences, and visits to contractor facilities. The Development Branch, DPD, the deputy director of plans, and the deputy director of plans was responsible for monitoring and reporting to the project director, Mr. Bissell. This relationship between customer and contractor enabled them to make decisions more quickly. The direct channel relationship saved valuable time as well as inducing the contractors to exert greater effort to achieve goals.

A-12 Camera Development.

A funded study by Perkin Elmer in 1959 explored the feasibility of developing a precision photographic reconnaissance sensor for use in a Mach 3.5, 80,000-foot altitude vehicle, having a photographic range of 2500 nautical miles.

The principal problem areas identified in the PE study were:

a. Control of the camera environment, I e thermal, pressure, and motion.

b. Effects of turbulence and shock waves on the optical wave front entering the camera. Optimizing the components of the sensor system provided photographic range, swath width, and resolution within the confines of the available space. It provided a window for the camera to look through that was 550° F on the outside and less than 150°F on the inside, features that did not distort the imagery to an unacceptable extent. During the ensuing development and test phases of the Type system, these identified areas proved to be the most troublesome as they still limited system performance.

The design philosophy employed by Perkin Elmer reflected a need for controlling the camera operating environment and in some areas extending state of the art to achieve the maximum performance. The result was a sophisticated design that gave excellent results if adequately maintained and serviced in the field. However, concern about reliability resulted in letting a contract to Eastman Kodak for the development of an alternate system easy to maintain and with high reliability, but not as good a resolution potential. Both cameras were flight tested in other vehicles before the availability of the A-12 and both achieved their design goals. When the delivery schedule of the vehicles forced a decision for the production quantities of anti-radar, the other, the CIA selected the Perkin Elmer camera by its demonstrated reliability and better ground resolution.

The Land panel concurred in the decision to procure eight cameras. Eastman Kodak built two cameras.

Hycon developed a third camera system for the A-12 program that it introduced to the OXCART inventory to serve as a backup for the Perkin Elmer camera in the event the Eastman camera was phased out. It was a framing camera as contrasted with the panoramic design of the other two, and followed the large scale, large format, and long focal length design philosophy, so successfully used for many years with the Hycon B camera in the U-2. These characteristics limited the amount of photographic coverage and required an amount of thermal control of the optical system that so far was not available in the A-12.

Chapter 2 - Electronic Warfare Systems.

One of the original design goals of the OXCART program was to produce an aircraft with a radar cross-section sufficiently reduced to make it difficult to detect and track with accuracy. The pursued anti-radar testing program was expected to provide the data necessary for optimum aerodynamic shaping and choice of materials to minimize radar reflectivity. No provision was made to equip the OXCART aircraft with defensive electronic systems in its, original, design. Its safety was its low level of detectability, its speed, and altitude.

The cross-section levels that the designers achieved were, in fact, quite low. However, the advances the Soviets were making in their radar defense network were equally impressive. It became apparent as more became known about the Soviet air defense system that the OXCART aircraft would be unable to overfly hostile territory without penetration aids.

The early warning TALLINN radar, which was replacing older equipment throughout the Soviet Bloc to detect and accurately track the A-12 during an overflight. There was evidence of improved performance in missile associated radar systems. A much more dangerous environment existed by the end of 1961 than planned against in 1959.

The new threat environment assessment had spurred an already intensive collection program to measure the parameters and sensitivities of new generation Soviet radars. Vulnerability assessments indicated that the SA 2 threat was sufficient to warrant the employment of ECM equipment and camouflage techniques. The resulting data and conclusions provided the base on which the CIA developed countermeasures for the A-12.

The conclusions were expressed in an OS report as follows:

(1) The Soviet TALLINN radar network can detect and track the OXCART vehicle at significant ranges. (Primarily line of sight)

(2) Soviet interceptors can be scrambled in time to attempt an intercept, however, given limited speed and altitude performance, as well as insufficient radar and missile capability, the probability of a successful engagement, was exceedingly small

(3) The US must consider the SA 2 missile system as a definite threat to the OXCART vehicle by the capability of the FAN SONG radar to achieve detection at an adequate range to permit the launch of at least one and probably three GUIDELINE missiles.

These statements led quite naturally to the requirement of exploring all means of reducing the vehicle's vulnerability. Consequently, the CIA actively pursued penetration aids and feasibility studies related to the following equipment concepts:

(1) FAN SONG radar detection and track indicator

(2) Strange, a signal detector whose purpose was to detect and identify radar signals that may emanate from missile guidance radars whose characteristics are presently unknown.

(3) Barrage and deception jamming of the FAN SONG radar and

(4) A missile launch indication radar. This equipment, correctly integrated, should provide the capability of the OXCART vehicle to monitor and react to the Soviet defenses in such a manner as to reduce vulnerability to an acceptable level.

Investigations of the various elements of the Soviet defenses continued to permit refinement of the present performance estimates. The CIA also critically examined typical missions to provide information about optimum techniques and tactics for successful penetration.

In the development of OXCART countermeasures, the CIA made a conscious effort to avoid techniques and equipment which duplicated, or adapted military hardware. This prevented compromise of military systems should the CIA lose an A-12 during an overflight. A family of new warning and jamming devices emerged under what was called the SUPERMARKET approach to the problem. It consisted of some electronic systems, passive and active, to warn the pilot of missile activity and to initiate jamming and confusion signals. The contractor, Sylvania designed redundancy into the total package to give a lower degree of vulnerability and to assure a high level of overall system reliability. The program made some combinations of systems available for use on a given mission. The CIA felt that judicious employment of the combinations would prolong the A-12's freedom of operation over denied areas.

The defensive electronic warfare system that Sylvania developed and employed in the OXCART aircraft as listed below with a brief functional description of each.

BIG BLAST (Active Jammer) Denied target range from the SA 2 radar to force the missile into a three-point guidance mode and early, arming of the fuse.

BLUE DOG (Active Jammer) Recognized missile guidance activity and actively transmitted false commands to the SA 2 missile guidance systems. PIN PEG (Passive DF System) passively intercepted SA 2 radar frequency signal. Located and positioned SA 2 radar site in azimuth within the vulnerable zone.

MAD MOTH (Active Jammer) Denied SA 2 tracking radar accurate angle information resulting to a great missile miss distances _ A. Signal Intercept Package collected ELINT data. System 6, an advanced ELINT collection system, which monitors the frequency range of 50 to 12,500 MH and provides an analog recording of the signals succeeded the MAD MOTH.

Significant Developments:

a. An air bar system of supporting the film during its passage through the camera resulted in extremely low friction, and tension on the film reduced abrasion and permitted right angle direction changes in the film travel

b. Vacuum windows with double glazing inhibited the transfer of high temperatures from the outside to the inside of the vehicle without distorting the camera's view of the ground. Studies and development in photographic window technology for supersonic airborne vehicles established manufacturing techniques and the effects of thermal transients on windows that proved invaluable in reconnaissance vehicles and sensor design

c. A method for synchronizing film velocity with an optical system scan rate permits continuous photography at vehicle rate and reduces dynamic camera motions. This technique has applied to other sensors in recent years. Over 200 test flights of the Perkin Elmer system in the years 1964 through 1966 established a degree of reliability and performance that culminated in 29 successful operational reconnaissance missions in Southeast Asia during the period from May 1967 to May 1968 with no failures. Each of these missions covered 100,000 square miles of denied territory.

The CIA removed the Eastman Kodak cameras from the inventory in July 1967 as part of the OXCART phaseout program.

The Hycon camera, having entered the program later, continued testing through 1967 and was validated as being operationally ready. The CIA never employed it on any operational missions due to the satisfactory performance of the Perkin Elmer camera.

Inertial Navigation System.

The CIA selected Minneapolis Honeywell Corporation (MH) to provide the inertial navigation system. It was a self-contained system and operates without the use of electromagnetic radiation or external references. In operation, it displayed present position, ground speed, and the direction and distance to go to any of 42 preselected positions. When operated in the appropriate mode the aircraft steered automatically to each point in the flight and planned sequentially with no pilot action required. The INS demonstrated a high degree of reliability, with a system specification error ratio of 1.5 nautical miles per hour of flight. The system consistently showed less than one-nautical-mile per hour error.

Automatic Flight Control.

The CIA selected Minneapolis-Honeywell Corporation to provide both the inertial navigation and automatic flight control system. The Firewel Corporation, a small company, operating in a basement designing regulators to convert coal stoves to kerosene became the prime source for oxygen life support equipment.

Minneapolis-Honeywell built the auto flight control system that included stability augmentation, autopilot, and air data system for the OXCART.

The three axis, stability augmentation system was a combination of electronic and hydraulic equipment which augmented the natural stability of the aircraft.

The system design rapidly took corrective action when its sensors detected pitch, yaw or roll motion that changed the aircraft's proper attitude. The resultant dampening effect was essential to maintain the stable platform necessary for high, quality photography. While the stability augmentation system, autopilot, and air data computers were not unique to the A-12, the aero-elastic and thermal problems associated with A-12 flight increased the system complexity regarding required response rates and cooling requirements.

Side Looking Radar.

The side-looking radar designed for the OXCART vehicle was a high-resolution synthetic aperture radar using an optical recording of the Doppler signal and using ground-based optical correlation techniques. The CIA procured two flight units and one engineering breadboard along with a strip correlator and two detail correlators for support.

The radar operated at a frequency of 9.432, gigahertz with a pulse width of 20 nanoseconds. Peak power was one megawatt, and the pulse repetition rate was 4,000 pulses per second.

The receiver noise figure was 62 dB (decibels). The system's resolution was 15-foot range and 8 feet in azimuth along the track with over a swath 20 nautical miles wide centered about 30 nautical miles off track. The 500 feet of thin based film used for original recording covered 1,000-miles of the track.

The Scientific Engineering Institute-proposed system weighted was about 950 pounds with an antenna length of 10 feet. The system, first proposed in April of 1960 saw the development of the equipment initiated in August 1960. Westinghouse was the prime contractor for the development of radar equipment, and Itek developed the airborne recorder and the optical correlators. The initial flight test started in March 1962 in a modified F-101 aircraft. In 1964, the system arrived at Area 51 and underwent extensive ground checks. Subsequently, in early 1965, the CIA returned the system to the factory, and F-101 tests continued for lack of OXCART aircraft availability because of other higher priority activities.

The F-101 VooDoo made over 200 test flights, and in late 1965, following a review of the requirement and usage of the OXCART radar, the CIA decided to reinstate the flight test program. The equipment returned to the field in the fall of 1966 for initiation of the flight tests that initially proved unsuccessful because of the antenna breaking down when operated at altitude. The manufacturer corrected the problem in December 1966, with the flight tests programmed to resume in early. 1967. However, there was only one aircraft configured for accepting the radar. Thus, it became a victim of the phase-down decision. An operational capability required to outfit additional aircraft with modifications extended downtime beyond the phase-down deadline.

The development of the radar designated as the AN/APQ 93 (XA 1) occurred under the guise of an air force secret procurement. The program, also known as INVAR, required additional work on the F-101 in support of Navy radar programs. The capability of the APQ 93 was better than most side looking radars but was somewhat lower in performance than some test and development models that came into existence.

OXCART Life Support Equipment.

At the time, Lockheed was developing the A-12 aircraft; there were no operational aircraft in the air force inventory with an equivalent performance envelope. Aircraft existing in 1959 such as the F-104, B-58, F-106, U-2, and X-15 all presented similar problems regarding life support requirements. Life support equipment used in such aircraft provided a necessary foundation to develop the life support equipment for the A-12. Life support equipment had, in general, two basic functions:

To enable the pilot to fly the aircraft and complete the mission throughout the entire performance envelope of the vehicle, under both normal and emergency operating conditions and

To enable the pilot to escape from the aircraft throughout safely, its performance envelope in the event of catastrophic emergencies. A-12 life support equipment, to provide these functions had to include a safe and reliable pressurization, air conditioning, and oxygen system with adequate redundancy and duration to meet any operational requirements. Also, the equipment had to provide the pilot with backup pressurization and an emergency oxygen supply, and protection from wind blast, high temperatures, deceleration and spinning if an emergency escape from a disabled aircraft.

Specific Problem Areas

The developing of the life support equipment for the A-12 presented specific parameters and problem areas:

(1) Pressure altitude - The A-12 operated up to a maximum altitude of 100,000 feet above sea level.

(2) Speed - The A-12 performed up to a top speed of Mach 3.5 at maximum, altitude, or a maximum equivalent airspeed of 450 KEAS,

(3) Range/Duration. - The A-12 had, because of aerial refueling, a worldwide range with flight durations more than 8 to 10 hours.

(4) Windblast/Deceleration. - The worst wind blast deceleration conditions under which a pilot could eject was Mach 3.2 at 74,000 feet, which exposed the pilots having to eject, to an instantaneous maximum dynamic pressure (wind blast) of 950 pounds per square foot and a maximum linear deceleration of 9 gs.

(5) Temperatures - Under normal A-12 operating conditions the pilot required protection from radiant heating due to aircraft skin and windshield temperatures approaching 400° F. An ejection at 100,000 feet and Mach 3.2 exposed the pilot to an instantaneous stagnation temperature of 800° F. Deceleration and free fall from high-altitudes exposed the pilot to temperatures of 67 F or lower,

(6) Spinning - If a pilot ejected at maximum altitude, where it was unsafe to open a standard parachute, he would enter a flat spin which would be uncontrollable. Such a turn, which could exceed 200 RPM would produce serious to fatal injuries: and would likely result in parachute failure when it automatically opened at a lower altitude. A-12 Life Support Equipment Development. Life support equipment falls into two categories: (1) aircraft systems, and (2) personal equipment.

The following paragraphs describe the A-12 equipment and briefly cover the development of each item. Lockheed Aircraft Company developed all aircraft systems and included the pressurization and air-conditioning systems. The cockpit pressurization schedule for the A-12 was unique as compared to existing high-performance aircraft.

The cockpit remained unpressurized from ground level up to a flight altitude of 26,000 feet. At flight levels from 26,000 to 100,000 feet, the cockpit remains isobaric at 26,000 feet (5.25 psia). This schedule gave:

(1) Optimum temperatures and airflows to the cockpit air conditioning system, especially during climb and descent

(2) Lower oxygen consumption consistent with the adequate physiological protection

(3) The safest pressure differential across the cockpit glass during critical refueling maneuvers.

The air conditioning system chosen was also unique in that an entirely dual system was selected. The unique problem of aerodynamic Keating of this vehicle dictated the provision of a complete redundancy of air conditioning for mission completion and adequate pilot protection.

Lockheed for mission completion and sufficient pilot protection. Lockheed, therefore, constructed a system providing cockpit/pilot cooling and a second system providing Q and E-bay cooling. In the case of failure of the system providing cockpit/pilot cooling, the pilot would operate a crossover selector to divert the Q and E-bay cooling to the cockpit/pilot. Changes and modifications made to the air conditioning system over the years continued until achieving peak efficiency, and maximum pilot comfort in 1966.

The Oxygen System

The critical nature of operational concepts required a redundant oxygen supply and delivery system in the A-12. The initial system consisted of two high-pressure oxygen cylinders and associated plumbing, reducers, and regulators. A failure of one system would not affect the pilot because the second system would continue to provide the pilot with his life-sustaining oxygen. Such a failure affected only the duration of the supply.

The calculated duration for both systems was 15 3/4 hours while the failure of one system after 5.4 hours of flight would reduce the total oxygen duration to 10 2 hours available. In the period 1964 65, the oxygen, systems converted to liquid oxygen (LOX), using two 10-liter converters, either one of which provided more than 14 hours of oxygen availability at altitude. In addition to the added safety margin, this change increased reliability and reduced weight and volume occupied by the oxygen system.

The Ejection System.

In 1959, Lockheed Aircraft Company analyzed the possible escape systems for the A-12 that included fuselage nose capsules, encapsulated seats, and rocket catapult seats. Lockheed determined the requirement for a pressure suit in any case, due to the potential loss of cockpit pressure. The most reliable, lightest, smallest, and safest escape system satisfying the high and low-level exit problems for the A-12 consisted of a full pressure suit and a rocket catapult ejection seat.

The pressure suit proposed was that developed for the X 15, because of its demonstrated capability for resistance to wind blast and elevated temperatures, which exceeded the strength required for use in the A-12.

The Upward Ejection C 2 Seat Also Used in the F 104.

The modified C 2 ejection seat pyrotechnics qualified for service use at the elevated temperatures encountered with the A-12 at 300° F versus 160° F for the F 104 pyrotechnics. The escape system received qualification through, testing including component tests, system functional (breadboard) tests, static tests, Edwards AFB track tests, aircraft drop tests at El Centro, California, and finally by in-flight ejection tests using a specially equipped F 106 at El Centro.

The test period covered a span from 1960 through final in-flight ejection tests in August.1964.

The A-12 ejection seat came equipped with a rocket catapult mounted on the rear of the seat to propel the seat upward on vertical guide rails during ejection. During ejection, the seat to aircraft attachments and pilot to seat attachments disconnected by a set of quick disconnect fittings. It had one for all electrical leads, one for the suit vent hose, one for the regular aircraft oxygen supply, and one for actuating the emergency oxygen supply.

The seat incorporated the following features

(1) A headrest for support and positioning of the pilot's head during ejection.

(2) A centrally located ejection D-ring which initiates the entire ejection sequence and precludes arm flailing after ejection.

(3) A shoulder harness and inertial reel lock assembly that secured the harness automatically during ejection, with the exertion of a 2 to 3 G-force in a forward direction, or manually when desired by the pilot.

(4) A pyrotechnic operated automatic opening seat belt and man seat separator. An airspeed sensor attached to the aircraft pilot static system and sensed aircraft airspeed at ejection. For aircraft airspeeds were below 250 KIAS (later changed to 290 KLAS), the sensor selected a 0.6-second delay located between the seat ejection and seat belt separator actuation. For airspeeds above 250 KIAS (subsequently changed to 290-300 KIAS), the sensor picked a 4.0-second delay. This variable delay assured that the pilot automatically separated from the seat as soon as possible, and within the limits of tolerable G forces for deceleration and parachute deployment.

(5) Knee guards which erected forward to protect the pilot's legs.

(6) An automatic-foot retraction and retention system which pulled the pilot's feet into stirrups and held them: until automatically released at the man/seat separation. The-foot retention cables automatically cut at the man/seat separation or manually released by pulling a headrest mounted D-ring for firing the cable cutters in the event of rapid ground egress.

(7) A canopy removal system tied, pyrotechnically to the ejection seat D-ring, and removable by an internal and external canopy jettison handle.

Personal Equipment.

Pressure Suit/Oxygen System.

The David Clark Company in Worcestershire, Massachusetts started in the textile business with the development of unique knitted materials for specialty undergarments and over time evolved to making aerospace and communications related products. The company pioneered the development of the pressure suits used by the Military and NASA for decades.

The company, working with the Firewel Co. developed a new pressure suit for use in the A-12, which represented the latest improvement over the then existing X 15 pressure suit and the air force Standard A/P 225 2 full pressure suit.

The suit provided a comfortable pressure atmosphere for the human body when the cockpit or out of the cockpit environment posed a danger to the suit wearer.

The human body operated very well at ambient atmospheric pressures up to 10000' above sea level. Sea level barometric pressure is approximately 14.7 PSI. The barometric pressure at 85,000' is about .25 PSI. The suit provided an artificial atmosphere for the pilot's body and body pressure at the same time it provided a pure oxygen atmosphere to the pilot in the helmet for breathing.

The S-901 pressure suit came equipped with redundant control regulators for safety. Under normal situations, the A-12's pressure ventilation systems provided the air and oxygen for the suit. In an emergency, backup supply enclosed in the parachute backpack supported the pilot after ejection from the plane.

In normal flight conditions, the airplane systems provided a ventilated (cooled) atmosphere with 100% oxygen breathing for the pilot. The helmet included a headset and externally adjustable microphone. The transparent glass pressure faceplate includes a heating element, and a tinted external visor (sunglasses) as part of the design.

The helmet attached to the pressure suit with a locking ring fitting and connected to the suit main body by a cable cinch-down mechanism to hold the helmet down when the suit inflated.

All the A-12 project pilots came into the program with some experience with flight using pressure suits. Those flying the F-101 wore a partial pressure suit that utilized physical compression of the suit garment to hold the body under pressure. Whereas, the full pressure suit used an enclosed body vessel that provided air pressure to simulate normal atmospheric conditions on the body. The pilots considered the partial pressure suit more of a "get me down safely" suit rather than a mission completion garment.

The David Clark Co.

Following their physiological exams by the CIA, the pilots went to the David Clark Company for tailored, suit fitting. During Project OXCART, the David Clark Company operated out of an old brown, stone building in the old textile district,

The tailors built the suits in stages, using some off-the-shelf parts. The suits started with "Long Johns" type cotton underwear that fit against the pilot's skin in white and olive drab colors. The two-piece underwear of standard size contained a turtleneck and extended sleeves that made the next part of the suit quite comfortable.

The inner pressure garment: part of the suit acted as a body glove. It was a flexible, waterproof fabric that inflated by air under pressure when needed. This garment is a waterproof flexible fabric.

The net link part of the suit provided support netting that overlay the inner pressure garment to prevent "blowouts" of the pressure garment. The suit builders considered this a critical part of the suit system and tailored to the person using the suit.

The outer protective garment, last part of the suit provided the scuffing and contact protection made of a heavier canvas-like material with sewn on pockets. It came in several colors. The OXCART pilots chose either white or silver.

The coverall provided the attachment points for the pressure regulators, ventilation fittings, flotation bladder, the glove connections, neck ring and the communication interfaces. The suit fitting took considerable time, first the building of the pieces, then a trial suit fitting followed by repeated adjustments until satisfied that all was right.

While assembling the suit, the helmet department made a casting is made of the pilot's head to obtain a properly fitting helmet for mating to the suit. Finally, the pilot donned the suit to check for basic comfort, leak testing, and readiness for the functional testing. The Firewel Corporation, a company other than David Clark Company, performed the function testing and training.

The gloves of the pressure suit provided a major part of the suit system. They attached to the suit by ring connectors, much like the ring connector used to attach the helmet to the upper body of the suit. They tailored the gloves to the user and somewhat similarly built like the pressure suit with a couple of layers, an inner pressure glove, covered with a protective outer contact surface. A soft, thin layer lined the palms of the gloves to allow good feel contact with switches, knobs, and the control stick. A material like the outer protective garment made up the outer (back of the hand) part of the glove.

The color of the gloves matched the outer garment but with the palm of the glove a black colored suede finish. The gloves contained "break-bars" in the palm of the glove and at the wrist for when the suit inflated along with the gloves. The glove became stiff, preventing the pilot closing his gloved hand to hold the stick grip. The break-bars are segmented metal rods that connected to lace, lanyard on the back of the glove. After suit inflation, the pilot could easily pull each of the cord straps to cinch up the segmented break-bars, thus making a crease in the glove palm and wrist.

A pocket on the left-hand glove provided for the pilot wearing of a wristwatch. The A-12 contained two panel-mounted clocks, but someone must have voiced a requirement for a watch pocket on the glove. Only the Bulova "Astronaut" battery powered watch could sustain reliable operation in the A-12 cockpit environment. It used a tuning fork time base and was impervious to the sometimes 140 degrees' Fahrenheit cockpit temperature. The point of interest, Frank Murray still wears his issued Bulova "Astronaut" watch at the Roadrunners Reunion.

David Clark Company provided boots and spurs, the boots oversize to cover the "footsies" of the pressure suit. All OXCART pilot suit boots came as white leather with a "quick-don" zipper in the front lace area.

The boots resembled combat boots, but of softer leather. Nobody walked very far in these boots.

The provided set of spurs made the pilot's legs compatible with the foot restraint feature of the Lockheed ejection seat. When the pilot entered the cockpit, he stood in front of the seat, then pushed his heels down in front of the seat base, engaging the spur notch into the cable ball fitting. Then he could sit down, moving his legs forward with the cables involved on the Spurs. The F-104 and Convair F-106s used this spur arrangement with a similar ejection system.

David Clark Company built an extra pressure suit for each pilot, one colored white, and one colored silver.

The suit testing occurred immediately following the pilot's suit manufacturer. Some of the David Clark Company technicians accompanied the pilot to the Firewel Corporation facilities in Batavia, New York for environmental testing/training.

The significant improvements and changes from previous suits were a dual oxygen regulator and dual suit pressure controller for increased reliability and safety. An increased helmet visor thickness prevented deformation or failure from the high temperatures encountered in the A-12. The outer cover of the pressure suit contained a flotation vest for water survival, and an aluminized flame: resistant fabric outer cover for protection from high cockpit temperatures or high stagnation temperatures encountered upon ejection at maximum altitude and airspeed. Parachute /Emergency Oxygen System Standard ejection seats, pressure suits and aircraft systems (oxygen, pressurization, and air-conditioning) required upgrading and improving for use in the A-12. The efforts involved represented product improvement. However, the personal parachute needed to complete the A-12 life support system was an entirely new concept and required considerable development and testing efforts.

The development and testing program started with a complete review of parachute systems in existence in early 1960 and an initial maimed parachute jump in September 1960 and ended with final configuration full system testing (parachute and ejection seat) by in-flight ejection from an F i06 during August 1964.

The first parachute jump in September 1960 made by Capt Harry Collins, the program manager and primary test subject throughout the development, established the requirement for an entirely new parachute concept. In this jump, the subject wore a full pressure suit, a packed survival (seat) kit, and a standard parachute.

The jump program allowed a 50-second free fall to determine the body position. After only 40 seconds of free fall, the subject spun at a rate of 140 to 167 RPM with the head and back in the down position.

The human jumper overrode the automatic timer and pulled the parachute, ripcord, and sustained high opening forces. Upon parachute opening, the spinning caused the parachute suspension lines to wind up approximately 28 times, creating the danger of chute collapse. This jump indicated that an A-12 pilot ejecting at maximum altitude required a stabilization parachute to keep him in an upright and stable position during free fall to an altitude, where the main parachute deployed (approximately 15,000 feet).

There were five model configurations developed and tested during the entire period. Models A, B, and C was purely developmental, and Model D was the first operational configuration. The provider made a total of 281 tests on these model parachutes and their subsystems during the period from January 1961 through August 1963. During. 1964, the provider developed and tested the fifth and current parachute configuration (Model E). This configuration design improved comfort, reliability, and maintenance of the parachute. An additional 37 tests were required to qualify this model testing being, completed in August 1964.

The personal parachute developed for and used in the A-12 had the following principal features:
(1) A 78-inch diameter, automatically deployed, ribbon type, Hemi SF lo stabilization parachute for

maintaining the pilot in an upright position during the descent from maximum altitude to 15,000 feet,

(2) A 35-foot diameter, main parachute canopy that automatically deployed at 15,000 feet, and

(3) An emergency oxygen system that automatically activated upon ejection consisting of two 45 cubic inch high-pressure oxygen cylinders integrated with the parachute back pan containing the automatic parachute actuators.

EG&G Tells the CIA to Take a Vacation

In 1958, EG&G, Inc. was borrowing its Atomic Energy Commission engineers and technicians to support the CIA's Area 51 Department of Defense activities. AEC operated with a "Q" level security clearance whereas the CIA and DOD used a Top-Secret level. On 9 July 1958, EG&G sought to temporarily close Area 51 for five weeks beginning 1 August because of an urgent temporary need for some of its Area 51-assigned personnel. EG&G suggested the CIA take a vacation.

Surprisingly, the CIA agreed to the shutdown as Project GUSTO was not due to complete until 1 August. The CIA facility did shut down with EG&G continuing with any maintenance work and Range improvements needed before the resumption of operations in September.

How the Support Contractor Companies Intertwined

By 1963, contractor EG&G was operating a DOD and CIA cleared group identified as special projects to bring expertise to the CIA as a permanent party operating the facility. EG&G special projects contracted with the CIA on a no-charge-for-use basis of all facilities provided by the Government provided such use did not interfere with the purpose of the facilities.

During the latter part of 1963, personnel problems at Area 51 and throughout the OXCART program continued to be a point of issue. Many of the support staff were nearing their regular rotation date, and there were no replacements available. Some had extended to June and July 1965 because their children were in school and they did not want to break-up the school years. Others did not want to sell their homes during the spring period. The CIA found it mandatory to start recruitment immediate to preclude a mass exodus without in-place transitioned replacements.

The CIA had contracted Lockheed Aircraft Corporation (LAC) as the prime contractor for the airframe, Pratt & Whitney Aircraft Division, United Aircraft Corporation, responsible for engine development and production, and project headquarters, DPD/DDP the direct contracting for certain auxiliary systems.

The CIA timely selected suppliers of cameras and other sensors, navigation, and flight control systems, life support and pilot environment equipment, and other services as needed. Perkin Elmer was the primary camera manufacturer; However, because of the extreme complexity of the design, the CIA decided to have Eastman Kodak build a backup camera system for the event the Perkin Elmer design ran into production problems. Minneapolis-Honeywell Corporation provided both the inertial navigation system and an automatic flight control system. The Firewel Corporation and David Clark Corporation proved pilot equipment and associated life support hardware.

In December 1963, the support of the EG&G special projects at Area 51 made it more indispensable when it gained access to the AEC's 1604 computer. With a half rack of equipment, the special projects could check the CIA's radar cross-section data in hours, saving the CIA having to transport it back to headquarters for reduction and analysis. The CIA agreed to the installation of the extra equipment and authorized the EG&G to use the computer facilities for the more routine work. The CIA would still do the data reduction and analysis in those cases where more advanced areas of work need EG&G's services, which at the time was putting equipment and programs together to simulate multiple targets against a SPOON REST radar. They expected to complete the data tasking and analysis by 1 March 1964.

The Development Branch, DPD, deputy director of Plans focused for the next two years mainly on the development, engineering, and manufacturing of the airframe, engine, and associated systems. Lockheed and of Pratt & Whitney experienced a magnitude of problems in terms both of dollars and of effort. The Lockheed designed aircraft system represented a significant milestone in aeronautical technology that ran into many challenges during production. The A-12's Mach 3.0 flight regime would generate extremely high temperatures on the airframe surfaces. During the design phase, the contractor evaluated such materials as steel honeycomb, high heat-treated steel, high-temperature aluminum and magnesium alloys, and titanium. The CIA and Lockheed chose a titanium alloy because of the high strength to weight ratio. Titanium retained its strength at elevated temperatures. Tooling costs were less than for steel honeycomb, which was the nearest material to it in characteristics.

Lockheed found tooling for titanium much harder than for conventional airframe fabrication. The company had to develop furnaces and treatments to handle the metal. There was only a limited supply of high-quality titanium metal, it was costly, and in low supply, to the few available suppliers filling the orders.

During the first year, the contractor, Lockheed, reported many problem areas. The amount of titanium and the high rejection rate of stock were the more severe problems because of poor quality control standards used by the metal suppliers. Other problems included time lost due to lengthy delivery delays, a high waste factor of the machining methods in addition to the rejection rate. An example of this occurred in the case that required machining down a 1,000-pound billet to an odd shaped 66-pound fitting. The machining costs ran three to four times higher than initial estimates.

To overcome these deficiencies, Lockheed developed metal handling techniques that considerably increased the costs and time involved.

On 14 September 1960, the contractor's revision of the delivery schedule slipped the first aircraft's delivery four months, changing the date of the first flight for Aircraft No. 1 from 1 May 1961 to 30 August 1961.

The technical problems encountered along the way changed the prices from the August 1959 quoted 12-aircraft production price of 96,000,000 to $103, 784,000 in January 1960.

The prime contractor's select subcontractors for various aircraft system components, parts, and materials experienced a myriad of problems with design and production encountering the inflexible demands imposed by the expected operating environment of their product.

The hitherto satisfactory design for parts and fittings now proved unsatisfactory in this new aircraft, requiring subcontractor research and development programs be initiated to find new materials and methods to meet the rigid specifications imposed. The result of this effort was much closer than usual contractor relationships and sharing of technical knowledge. Suddenly, Lockheed was collaborating with North American who was developing the X-15 rocket ship and Mach 3 XB-70 bomber. Even with increased costs, the effort produced incalculable benefits nonetheless. The CIA's OXCART program was destined to bear the financial brunt for many advances in the science of, supersonic aerodynamics which government and industry would profitably employ in future programs.

Engine Problems

In August 1960, P&W informed project headquarters it was experiencing considerable increases in cost, particularly the materials in the development and production portions of the contract. Development costs increased by 12 million and the production costs for 36 engines by 10+ million. Project headquarters, while greatly disturbed, agreed to fund the additional development costs within limits. It was forced to reduce the order for engines from 36 to 30 because of the limitation on funds available. Pratt & Whitney gave assurances that it could accomplish the J58 program within the revised dollar estimates.

The rapidly increasing costs of the engine program were due to various problems that surfaced as the development phase proceeded. The high Mach performance design introduced temperature environments that were never previously experienced or anticipated. Taking the engine thrust requirement, for example, it dictated a turbine design to handle an average inlet temperature of 1900° F.

Early tests of the turbine on the stand revealed hot areas with extreme temperatures peaking at 2500° F. The combustion section and turbine fabrication materials used could not sustain such an extreme condition.

Pratt & Whitney spent numerous costly engineering manhours in the redesign of the combustion section and turbine, having to select new materials and alloys for the turbine vanes, and blades that required verification on the test stand each change incorporated.

Change or modification in one area of the engine necessitated redesign rework, experiment, and test on other engine parts. Redesigning the burner cans and diffuser cases evened out the temperature profile for enduring the extreme heats.

Each time the company introduced new materials, it caused fabrication difficulties that required devising new and very expensive welding and machining processes. The light afterburner liner sections buckled during test runs, tearing loose the buckled sections, and blowing out the engine tailpipe. A structurally inadequate compressor rotor design required reconfiguration.

Major engine components of the Vickers hydraulic pumps and Hamilton Standard main and afterburner fuel controls malfunctioned far ahead of their expected operating times, and again, the heat played as the major contributor to the durability of materials and equipment failures. The engine component suppliers found themselves having to go into crash engineering development and testing to find satisfactory designs and materials with which to modify and retrofit their hardware which was already on production lines.

Each new fix to the engine or a component added weight that became an accompanying penalty. A vicious cycle developed where for each problem that arose, a costly and time-consuming domino effect resulted.

Pratt & Whitney dropped development of the J58 from a Mach 3.2 engine to a Mach 3.5 capability in December 1960 when weight and fuel tradeoffs resulted in no significant improved performance for the A-12 mission.

Selecting Area 51 For Continuing Development

In the fall of 1959, the Development Projects Division took up the matter of choosing a domestic flight test and operational site. The OXCART operational concept envisioned flying missions from the site and return with range extension accomplished by aerial refueling, en route.

The strong security requirement eliminated active military bases that placed the project subject to public scrutiny. The criteria required a venue remote from metropolitan areas, civil and military airways subject to aerial observation or collision. The CIA needed a site easily accessible by air, having good year-round weather, and capable of accommodating large numbers of personnel. The facility required a POL storage facility relatively close to an air force installation, with a nearby source of labor. Above all, the project required an 8,000-foot runway.

The CIA inspected and evaluated ten Air Force bases programmed for closure. The CIA found that all failed the remoteness test, and had unacceptable annual operating costs. The CIA considered only Edwards Air Force Base and the Watertown site worthy of serious consideration.

The CIA questioned the security at Edwards AFB, whereas the security Watertown provided by its location within the restricted AEC reservation made it ideal for clandestine, aircraft testing and operations. However, Watertown was deficient in personnel accommodations, POL storage and its runway inadequate. What it lacked in the physical plant could be rectified with a relatively inexpensive construction program.

On 23 December 1959, the DDP approved Watertown as the primary domestic, the base for Project OXCART, subject to the concurrence of the air force and the AEC. The decision, however, did not imply the start of a significant program of base improvements and new construction. Until given a final go-ahead to build aircraft, The CIA limited any expenditures to improve Watertown facilities to those necessary to sustain the air force test program.

The following month, the CIA approved the full-scale program. This authorization was where the planning in advance for the selecting of an operating base paid off. The early identification of the Watertown base permitted headquarters engineers to develop an orderly plan to ready the base for the aircraft and associated, ground equipment on schedule.

The CIA consulted with Lockheed from the very beginning on the required facilities to support the flight test program. Lockheed estimated major items such as runway loading weights, monthly fuel consumption, hangars and shop space, and numbers of people necessary to conduct the flight test. The headquarters OXCART department consulted with other contractors on their needs, and armed with the major requirements, came up with a preliminary construction engineering plan.

Since the Watertown base was within the AEC nuclear test site, it was necessary to obtain AEC approval to reactivate the base. An agreement drawn up in 1955 between the CIA and AEC allowed the CIA to use the Watertown site for the U-2 flight testing and training.

To preserve cover and security, the AEC ostensibly managed the facility. An AEC proprietary concern had performed all construction, maintenance, and housekeeping, the Reynolds Electric and Engineering Company (REECO).

The CIA renewed this same type arrangement with the AEC and REECO. A considerable advantage to having REECO perform the work, was it being a qualified construction organization equipped to handle the job and was that it was already on the scene. Its relation to AEC provided the cover essential to the security of the operation. Its employees already possessed the Atomic Energy Commission "Q" security clearances to work on the Nuclear Test Site. The AEC Q clearance equated to the CIA and Department of Defense top-secret clearance.

A cover story for the reactivation of Watertown was prepared in the form of a public information release to use in reply to press or other inquiries. It stated that the AEC facilities at the site had been made available to EG&G to conduct radar studies with support from the air force.

The remote site was chosen to reduce the likelihood of outside interference affecting instrument calibration and ultimate test results. Use of AEC facilities did not affiliate the AEC or any of its programs with the work going on at Watertown. The CIA furnished the AEC and EG&G with the text of the press statement to use as required and instructed them to pass further requests for information to Headquarters, Air Force, Office of Information Services, who would, in turn, seek replies from project headquarters.

Chapter 3 - Project 51

The code name "Project 51," which some claim as being the genesis of the Area 51 name, identified the phase where Lockheed planners estimated cost requirements for monthly fuel consumption, hangars, maintenance facilities, housing, and runway specifications. Project 51 enabled the CIA to produce a plan for construction and engineering to facilitate developing America's first stealth plane, a plane to replace the U-2.

A CIA cover story stated the facilities are preparing for radar studies conducted by an engineering firm with the air force support. Reynolds Electrical and Engineering Company (REECo) performed the construction at the site for project 51. Workers ferried in from Burbank, and Las Vegas on C-54 aircraft for base construction began on 1 September 1960 and continued a double-shift schedule until 1 June 1964.

The A-12 required a runway at least 8,500 feet long, requiring the pouring of about 25,000 cubic yards of concrete.

A 10,000-foot asphalt extension, for emergency use, cut diagonally across the southwest corner of the lakebed. Construction contractor, Holmes & Narver, Inc. (H&N) constructed a special pylon on a paved loop road on the western side of the lakebed. The company then built a new airstrip to replace the existing 5,000-foot runway found incapable of supporting the weight of the A-12.

An Archimedes curve approximately two miles across was marked on the dry lake, allowing an A-12 pilot that is approaching the end of the overrun to abort to the playa instead of plunging the aircraft into the sagebrush. Area 51 pilots called it "The Hook."

They marked two unpaved airstrips (runways 9/27 and 03/21) on the dry lakebed for crosswind landings. Kelly Johnson felt reluctant to construct a standard air force runway, with expansion joints every 25-feet, fearing the joints causing unwanted vibrations in the OXCART A-12 aircraft.

At his suggestion, they constructed the 150-foot-wide runway in segments; each made up of six 25-foot-wide longitudinal sections. They staggered the 150-foot-long sections to put most of the expansion joints parallel to the direction of aircraft roll to reduce the frequency of the joints. They, the CIA, Lockheed, REECO, and EG&G completed all necessary facilities by August 1961.

An EG&G subcontractor, Reynolds Electrical, and Engineering Company (REECo) obtained, dismantled, and erected three surplus Navy hangars on the base's north side, designating them as Hangar 4, 5, and 6. They built a fourth, Hangar 7, and transported and made ready for occupancy more than 130 navy surplus Babbitt duplex housing units to the base.

REECo converted the original U-2 hangars to maintenance and Machine shops. Facilities in the main cantonment area included workshops and buildings for storage and administration, a commissary, control tower, fire station, and housing.

Understanding the Work Force

For the reader to appreciate the task of preparing, the base for occupancy, certain facts are necessary. Due to its location, 120-miles from the nearest metropolitan area (Las Vegas), personnel were required to live at the site during the work week. In 1960, the number of people involved with reopening Area 51 started with 75 and grew to 150 by the year's end. This figure represented chiefly contractor personnel engaged in the anti-radar testing and base construction.

Except for the EG&G and REECO contractors headquartered in Las Vegas, no contractor personnel could maintain residences in Las Vegas. Lockheed supplied a C-47 shuttle service between Burbank and Watertown for its freight and passenger needs. A chartered D-18 (Lodestar) was provided for transportation between Las Vegas and the base, chiefly to support the EG&G contingent. Surface access to Watertown was quite difficult because of distance and the fact that the only road leading into the site had deteriorated since 1957.

Immediate needs at Watertown were fulfilled on an as needed basis during 1960. The CIA procured trailers as surplus housing to billet construction workers as they arrived. REECo employee, Ernest Williams dug a new well. The CIA provided limited recreational facilities and assigned an Agency staff civil engineer to the base to provide the necessary engineering guidance to the Chief of Base. The engineer also performed liaison for project headquarters with the construction contractors.

By May 1960, the Operations Branch, DPD, deputy director of Plans, estimated a requirement for 500,000 gallons of aircraft fuel per month. Neither storage facilities at, nor means of transport to Watertown existed. The refinery carried fuel by rail to Las Vegas. After examining airlift, pipeline, railroad, highway, and combinations of these modes of transport, the

Kelly Johnson
Lockheed Skunk Works

CIA determined that truck transportation was most economical. The decision meant resurfacing eighteen miles of road leading into the facility to bear the weight of the vehicles, but this was cheaper than constructing rail or pipeline systems.

Initial estimates, of A-12 runway requirements, called for an 8,500-foot length, which meant lengthening the existing 5,000-foot asphalt strip incapable of supporting the weights of the A-12. Plans for a new concrete runway were drawn up, and preliminary engineering began.

The Watertown location within the section of the AEC Nevada Test Site was officially designated "Area 51" on AEC maps. The designation afforded the CIA's facility the standard security protection as the rest of the AEC reservation though some 14,000 acres of the land that the AEC sectioned at Area 51 lay immediately adjacent to and overlooked the Watertown facility and was not part of the AEC complex. Unrestricted access to the adjoining property would permit unauthorized viewing of the base's activities. Legal measures were instituted to have the land withdrawn from public use to provide a protective screen.

In October 1960, the CIA, and the air force established the Detachment 1, 1129th SAS (The Air Force) Special Activity Squadron, Mercury, Nevada as an air force cover organization to give the activities at Watertown a legitimate character. Its parent unit was a fictitious unit at Fort Myer, Virginia. Administratively, the squadron was at Bolling AFB. All Agency staff and contract personnel received the air force documentation upon assignment to the field activity.

Constructing Area 51 Runway

1129th SAS enlisted crew for the F-101 VooDoo chase planes

Arriving workers at Watertown lining up to pass through security. *CIA via TD Barnes Collection.*

Note the damaged TV antenna from a sandstorm and the guard climbing the water tower to his guard shack. *CIA via TD Barnes Collection.*

The latest weather maps at Watertown. *CIA via TD Barnes Collection.*

Chapter 4 - Area 51 Cadre

For cadre to operate the facility, the CIA's Area 51 station preferred men married with two children and common interests. The CIA felt this created the necessary bonding for sharing national security concerns. It did. It developed a cohesiveness where they worked together all week and then played together at the lake or on the mountain during the weekends. Rarely did they talk shop, and even under these conditions, they never did so with the spouses or children present.

Each member of the special projects team and their families had undergone the same security and compatibility evaluations, though many of them evolved from the atomic testing side. Consequently, almost every member the special project team carried both an AEC Q security clearance and a DOD top-secret security clearance.

The CIA selected each member of the special projects team for his and her unique qualifications, family stability, ethics, integrity, and moral qualities as well. (Denise Haen based in the Las Vegas office.)

Almost all engineers in the special projects team chose a senior technician classification, allowing them to escape the "salaried" pay status of an engineer. As senior technicians, the special projects team drew almost the same pay as engineers. However, after the first eight hours, their pay rate increased to time and a half for four hours. Their pay scale then went to double time straight through, 24 hours a day until they arrived back in Las Vegas on Friday evening.

They received free food and lodging and even provided with snake proof boots for when they worked at the pylon out on the lakebed.

Area 51 deputy commander Werner Weiss and his successor, Dick Sampson of the CIA, accommodated the permanent party personnel at Area 51 with a small BX. It stocked snacks and various personal hygiene items. Other amenities included a swimming pool, exercise room, softball diamond, putting green, pool room, and a three-stool bar called "Sam's Place," named after Dick Sampson.

The amenities at Area 51 reduced the hardship on the special projects team members maintaining two residents. Area 51 provided each room in a row of Babbitt housing units. Each house contained a small living room where the team played poker and watched 8 mm movies played on a film projector.

The special projects group usually banded in two groups for housing, one group of the boating enthusiasts with boats moored on Lake Mead and the other as the Mt. Charleston cabin dwellers. The CIA emphasized personnel selection including common interests, not only with the workers, however, their families as well.

When Area 51 was reactivated in late 1959 to begin flight testing of the A-12, the site was administered by the 1129th Special Activities Squadron with Colonel Robert Holbury in command. His deputy was a senior Organization representative, Werner Weiss, a GS-15 with the Central Intelligence Agency. The CIA facility was composed of US Air Force personnel, civilian staff, and contract personnel related to the A-12 project, and the EG&G special projects contractor personnel engaged in CIA project research, the latter reporting to the agency's science and technology division at Langley.

Reynolds Electrical and Engineering Company (REECo), under the direction of the CIA management, performed site maintenance and housekeeping and supplied the facility with everyday housekeeping and maintenance supplies and equipment under the general administration of the Area 51 Base Commander.

The CIA station, Area 51, was located approximately 120 miles northwest of Las Vegas and s53 miles north of Mercury, Nevada, at the edge of the Nevada Test Site. It could be reached only by automobile and aircraft., Constellation from Los Angeles and C-47 from Las Vegas. Since there was no public surface transportation between Area 51 and Las Vegas, and since space on the contractor C-47 was available only on a first-come, first-served basis, personnel were urged to bring their automobiles since it may be their only means of transportation between the Area and Las Vegas.

The cost of living in Las Vegas was higher than the national average, probably one of the highest consumer indexes in the US. All personnel assigned to Area 51, whether they had a family in Las Vegas or another city, were assigned quarters and provided meals without cost during their stays on Base. Personnel lived in Babbitt houses or trailers, one or two men to a room. All homes and trailers were well furnished and had air-coolers and heaters. Towels and bedding, including linens and blankets, were furnished without charge. Incoming personnel was advised to expect only simple field living conditions with no locked storage suitable for the protection of valuables within the Area. Therefore, personnel provided their footlockers.

Television sets were available in all houses and trailers and personnel were requested not to bring their personal TV sets to the Area since antenna arrangements did not provide for unlimited hook-ups. Also, AM radio reception was poor during daylight hours, and FM reception was nonexistent. Those considering a transistor-type radio were told to consider no less than a 7-transistor unit.

The messing facilities operated by Reynolds Electrical and Engineering Company were excellent. The newly constructed mess hall managed by Murphy Green accommodated approximately 1,500 people, and hours of operation assured all personnel, regardless of working shifts, three full meals daily. There were no shortages, and fresh fruit, milk, coffee, tea, pies and cakes, salads and ice cream were always on the serving line.

Oblique looking south

The ordinarily extremely dry weather changed during the monsoon season of severe rainstorms and occasional snowstorms. The workers anticipated extreme seasonal temperatures with below freezing temperatures during the short winter months and extreme heat and intense sun during the summer months.

The facility had a free do-it-yourself laundromat, with soap and bleach furnished. Laundry could be sent to a Las Vegas laundry through REECo on Tuesdays and Thursdays weekly and picked up a week from the day sent. Dry cleaning laundry was expensive, for example, it cost about 85 cents to have a pair of trousers dry cleaned and about 45 cents to have a shirt laundered.

Informal wear was the order of the day during working hours. Walking shorts and T-shirts or sports shirts were the usual attire during the summer months. With few exceptions, all buildings were air-cooled or air-conditioned. The temperature reached 120 degrees outside.

For medical facilities, a United States Air Force staffed dispensary provided 24-hour medical coverage for the entire facility. The facility had no hospital facilities, so military personnel was authorized to use Nellis AFB facilities in Las Vegas while civilian personnel had to use hospital facilities in the Las Vegas or Los Angeles area. Nellis AFB provided a clinic for military personnel and their dependents assigned to the detachment every Friday by the Air Force Officer General Practitioner assigned to this station.

For recreation, the stationed a newly constructed conditioning building containing six bowling alleys installed by Brunswick Corporation, a dry heat room, enclosed swimming pool, handball court, squash court, physical improvement room, paperback library, and a basketball court. The bowling alleys had automatic pinsetters and ball returns, and charged a fee of 25 cents. The higher-grade movies charged a little extra.

The recreation hall contained a beer bar, hardback library, six pool tables, a TV room, hobby room, shuffleboard, and a snooker table.

The full-size gymnasium contained a basketball court, badminton court, and an exercise area. Outdoors was a softball field, four tennis courts, a basketball court, a volleyball court, a badminton court, a hand court, and a gold cage. CIA pilot Frank Murray and others took up a model plane flying, and many of those at the station later became Ham radio operators.

All privately owned autos required registering with the Area 51 security office. Auto insurance rates were extremely high in Nevada.

For this staying at the station over weekends, Camp Mercury and Indian Springs AFB held religious services with transportation available to anyone desiring to attend the service of their choice.

Barbershop services were available at Camp Mercury, Las Vegas, and Nellis AFB. Every Wednesday, there was a barber at Area 51 from 0800 to 1800 hours. The average price of a haircut throughout southern Nevada was $2.50. However, military personnel could get haircuts at Nellis or Indian Springs AFB for $1.25.

The station offered a small on-site PX in the recreation hall, a source for toilet articles, tobacco, magazines and newspapers, soft drinks, and beer.

There were no public telephones available on site for personal calls. The Area 51 military commander authorized any emergency calls from on-site.

For mail service, those at the station used the first-class post office at Camp Mercury. Civilians residing off-site could use P.O. Box 732, Las Vegas, and military personnel could us P.O. Box 882, Las Vegas. Both received daily service and mail delivery to the Area.

No animals or pets were allowed at the station. The workers could bring personal cameras to the Area. However, the Nevada Test Site and Area 51 prohibited any photography. Upon arrival, the owners registered all cameras with the Area 51 security office. The Nevada Test Site prohibited firearms of any type, However, the Area 51 workers could transport firearms to the Area provided they immediately registered them with the Area 51 security office.

Lizards, copperhead and king snakes, deer, mountain sheep, jackrabbits, bobcats, and badgers amply inhabited the surrounding desert area.

With Project OXCART, the military did not arrive until near four years into the program, when the A-12 was ready to fly.

No women worked on site. To that extent, initially, the air force even denied concurrent dependent travel for the air force support personnel assigned to Area 51, classifying it as an unaccompanied tour. Those in the 1955-56 U-2 program found the no dependent policy acceptable because all concerned knew they would leave once the training stopped.

The thought of spouses deploying to Las Vegas without their families drew such an outcry among the dependents the air force gave in and allowed the families concurrent travel.

It concerned the CIA that military families are suddenly showing up, their presence potentially exposing the secrecy surrounding Area 51. Consequently, the air force assigned its personnel to March AFB instead of Las Vegas, forcing those working at Area 51 to commute.

This concern did not apply to the special projects team that the CIA considered as a permanent party or cadre. Another reason was the CIA vetting the families of all personnel in special projects, screening that resulted in families used to being separated, used to no discussion about work, and families with common interests.

An interesting note on this is how the families stayed in touch for the rest of their lives; the children stayed in touch even if they attended different schools. An interesting statistic was the few numbers of divorces if any. Credit for this goes to the CIA's selection and vetting. Another factor was almost all of those chosen for special projects either had a boat moored on Lake Mead for owned a cabin on Mount Charleston, common interests that developed groups who worked together and played together with lifelong bonding.

Helen Kleyla, longtime secretary to CIA Deputy Director Richard "Dick" Bissell once accompanied Bissell to Area 51. On another occasion, Denise Haen, EG&G's contracting officer at headquarters came to the site for an inspection.

Housed and transported separate from any other tenants, members of the special projects team reported only to their customer. Many of those serving and unique to Project OXCART left Area 51 at the same time as their project. Named below are those CIA EG&G special projects cadre at Area 51 who remained for the MiG exploitation projects and identified as an exploitation team recognized as Organization #6300.

The addition of new highly classified projects tightened the compartmentalization and need-to-know environment within the special projects exploitation team. With Project OXCART, the special projects team worked on individual projects and for entirely different purposes. The arrival of the MiGs changed the focus of the special projects team from technical support to personnel having the experience and expertise to exploit Soviet technology along with supporting projects needing individual expertise offered by the team.

The special projects team members not talking about their work with each other did not mean they did not want to get to know each other. Working in such secrecy bonded the families. None felt comfortable socializing outside their team. They did not trust strangers. The result for the men and their families was them isolating from the outside world and bonding as a family.

Most understood in basic terms what the other did from giving them a hand with something. Nonetheless, they did not ask what, why, who, or how if they lacked the need to know.

The isolating did not mean the team members lost their curiosity. Quite the contrary, however, each member, with his specialized ambitions and professional goals, expected protection from snooping regarding his projects. The members assisted each other as needed. They, however, totally respected the proprietary rights of each other's thoughts and achievements.

These highly motivated workers did not work by the time-clock. The CIA selected each based on his pristine ethics, his needed specialty, and his drive to achieve the impossible. Those selected did not consider what they did at Area 51 as work — they felt it was a duty, privilege, and pleasure.

John Grace, a member of America's Who's Who obtained many of the radar systems in the RATSCAT radar array. Denise Haen handled the administrative needs for recruiting, security, and safety human resources from the EG&G building on Sunset Road in Las Vegas under Burt Barrett. Denise also held the distinction of being the first female to hold the title of Director of special projects Administration. Denise later retired from special projects after 31 years of service.

Dave Haen specialized in telemetry during Project OXCART. During project HAVE DRILL, he became involved with a new radar system. As with most of the special projects team, from that point on, his career remains classified top secret. Over the course of over 30 years in special projects, Dave advanced to Director of Site Support Operations. Dave married Denise, EG&G's contracting officer previously mentioned as having inspected the facility at Area 51.

Jim Freedman came from the atmospheric testing of the atomic bomb in the various remote areas of the world to the special projects team as an administrator and courier. Declassification revealed that each evening Freedman dropped by CIA commander Werner Weiss's office to pick up the dispatch to Langley. He dropped it off to a United Airlines employee at McCarran International Airport in Las Vegas, and the next morning repeated the procedure in reverse, delivering the dispatch from Langley to Werner Weiss at Area 51.

The termination of CIA Project OXCART in early 1968 drastically reduced the number of special projects personnel. The RIF took away key personnel such as Harry Phiffer, the one in charge of the engineers at the site. Wayne Pendleton flight-systems (flight) manager went to work for Lockheed, and Jim Tarver G-systems (ground) manager took an assignment elsewhere. Jules Kabat, responsibility for the two radar systems in the antenna building, the S-band radar with the 60-foot dish and a navy radar left Area 51 to accept an assignment in Vietnam.

Frank Harris operated and maintained the C-band radar. He and Bill McCloud, the antenna building lead tech, reported to Howard Schmit, the F-systems senior technical supervisor, and George Percy, senior technical supervisor for G-systems.

Credit goes to the previous U-2 and A-12 special projects teams for having in place this facility to conduct radar cross-section evaluations. Others included:

Dick Lampier- (G-systems engineer),
Stan Busby- (G-systems engineer),
Carl Newmiller-(Draftsman),
Thornton D. Barnes- (Nike Radar),
Dick McEwen-(G-systems),
Cowan Dawson- (C-band Radar),
Dick Wilson-(Q-bay),
Robert Pezzini- (Antenna building),
Vern Williamson-(PPA),
Dave Haen-(Q-bay),
Sam Gamble-(Draftsman),
Jim Freedman-(Admin),
? Helbert-(G-systems),
Jim Cates-(G-systems),
Eddie ?- (Admin),
Rocky ?-(G-systems Tech Supervisor),
McGlothen-(Clothesline),

The Carco C-47 Pilots Out of Albuquerque, New Mexico:
Roy Kemp-Chief Pilot,
Tom Hall,
Joe Cotton,
Hugh Starcher,
Tom Losh, and
Flo Deluna-aircraft mechanic.

The Central Intelligence Agency operated the Area 51 facility, yet, few know outside the special projects team of the CIA affiliation as cadre at Area 51. Even within the special projects team, the team members did not necessarily know what brought their contemporaries to Area 51. For illustration, the author will use his (Barnes') personal circumstance.

Sixty-five miles west of the Area 51 facility, Barnes conducted the Seven Sisters tracking for the CIA at Beatty. Only Barnes had the required security clearance from his previous involvement with the CIA related to Project OXCART, the A-12 reconnaissance plane.

He graduated after attending over two years of schooling at the Nike Ajax and Nike Hercules surface-to-air radar and missile maintenance schools at Nike Ajax/Hercules: USARADSCHFTBliss. He graduated from the six-month HAWK surface-to-air missile maintenance school at Hawk: USARADSchFBliss. This specialized education provided him with the unique radar experience needed at Area 51 for airborne high-speed RCS radar cross-section evaluations of the America's first stealth plane, the A-12. He brought with him several years' experience in tracking 3,000 mph missiles. He was one of only half a dozen experienced in tracking the Hypersonic Mach 6.7 X-15 rocket plane using NASA's SCR-584 Mod-2 radar at Beatty, Nevada

His job at the NASA radar site included other systems besides the radar, systems such as telemetry, DTS (data acquisition and transmission), analog to digital conversion, timing, and communications. NASA's overt activities did not require a security clearance.

Barnes still had his security clearance from his US Army service. Having it uniquely qualified him to serve the CIA need that required, on occasion, covertly using the NASA radar system for its ability to record a plane's velocity. Thus, he became a member of the Seven Sisters, a classified element of CIA Project OXCART and the CIA/Air Force Project KEDLOCK.

For these mystery flights, the NASA monitor at Beatty, Mr. Bill Houck, received a call requesting special track of an unidentified plane with the only recorded data being the station's velocity recorder. For these flights, he assumed the duty of radar operator and Houck provided the security to ensure other personnel at the station did not observe the track or data requiring special handling.

Air Force SAGE ADC (Air Defense Command) operated six of the radar systems. The radar at Beatty was an SCR-584 (short for Signal Corps Radio # 584) microwave radar developed by the MIT Radiation Laboratory during World War II. It replaced the earlier and much more sophisticated SCR-268 as the US Army's primary antiaircraft gun laying system

The CIA embedded him into the Area 51 special projects team to conduct airborne RCS stealth evaluations of the A-12 and later the MiGs of Area 51 for the combination of his unique qualifications and availability.

For airborne RCS tests of the A-12, the Lockheed flight test director called the EG&G special projects team for mission briefs.

As mentioned earlier, these training flights typically flew to Mountain Home, Idaho and back, passing overhead before landing.

At first, the special projects team found it difficult to locate and lock onto the plane traveling inbound at Mach 3. The radar operators searched for the return of the aircraft by scanning the horizon with the Nike radar used to track it on the inbound leg. Once the target appeared, the operator had only seconds to lock the radar onto the target. Heading inbound at Mach 3, the range decreased, and the elevation increased at too high a rate for the radar to slew and catch up.

The Nike radar operator acquired target acquisition by sweeping his beam at the expected azimuth for the target to pop over the horizon. With the radar azimuth and elevation locked on the target, the target's transponder answered with a spike for the operator to capture and lock in the range gate.

Once the Nike radar acquired a target lock, it drove the bull gear to align all other antennas on the target.

John Grace solved the target acquisition problem by bringing Barnes in from the NASA High Range at the Beatty Radar Station. Barnes, a Hypersonic Flight Support Specialist, routinely locked onto the much faster X-15 and tracked it as it passed directly over his radar site at Beatty. (On the NASA, High Range in Nevada, Barnes maintained his proficiency with the radar system, the data transmission, communications, microwave, and telemetry systems. He participated in tracking the Mach 3 A-12, YF-12, and M-21 Blackbirds, the Mach 3 XB-70, and the Mach 6.7 X-15.)

Once the A-12s deployed to Kadena for Operation BLACK SHIELD, the range shut down in June 1967 and remained so until early in 1968 when the MIG-21 arrived for exploitation.

Working at Area 51.

Almost all engineers in the special projects team chose a senior technician classification, allowing them to escape the "salaried" pay status of an engineer. As senior technicians, the special projects team drew almost the same pay as engineers. However, after the first eight hours, their pay rate increased to time and a half for four hours. Their pay scale then went to double time straight through, 24 hours a day until they arrived back in Las Vegas on Friday evening.

They received free food and lodging and even provided with snake proof boots for when they worked at the pylon out on the lakebed.

Area 51 Deputy Commander Werner Weiss and his successor, Dick Sampson of the CIA, accommodated the permanent party personnel at Area 51 with a small BX. It stocked snacks and various personal hygiene items. Other amenities included a swimming pool, exercise room, softball diamond, putting green, pool room, and a three-stool bar called "Sam's Place," named after Dick Sampson.

The amenities at Area 51 reduced the hardship on the special projects team members maintaining two residents. Area 51 provided each room in a row of Babbitt housing units. Each house contained a small living room with linoleum floors where the team played poker and watched 8 mm movies played on a film projector.

Some of the OXCART people were lucky enough to officiate at the PGA Tournament of Champions hosted by The Desert Inn. They took turns escorting golfers from tee to green. Capt Charlie Trapp walked with Nicholas, Palmer, Casper, Rodriquez, Snead, and others. He was within a few feet of them for every shot and comment.

SSgt Lin Kelly set all this up with a colonel that worked at the Atomic Test Site for the Area 51 golfers to play free golf anytime at The Desert Inn, Tropicana, and Sahara courses. Denny Sullivan and Trapp had many very competitive rounds of golf together (squash court also).

The special projects group usually banded in two groups for housing, one group of the boating enthusiasts with boats moored on Lake Mead and the other as the Mt. Charleston cabin dwellers. The CIA emphasized personnel selection including common interests, not only with the workers, however, their families as well.

Further grouping often occurred to segregate the team, per project or customer. Extracurricular activities such as poker, rental movies, reading, and so forth, developed, even more, sub-grouping per interests. (Note about these recreation activities: Lake Mead is on the Colorado River and the largest reservoir in the United States. Mount Charleston, officially named Charleston Peak, altitude 11,916 feet, is located about 35 miles northwest of Las Vegas),

CIA and other personnel staying at Area 51, such as Air Force, Lockheed, Hughes, Pratt and Whitney housed in similar houses, however, clustered apart from the others. There was very little association existed even in the special projects team outside one's professional group.

The author's radar on the NASA High Range is the same as the WWII Lend-Lease systems used by the Soviet Union against the U-2 and A-12 surveillance planes

The Special Projects RCS site.

For airborne RCS tests of the A-12, the Lockheed Flight Test Director called the EG&G special projects team for mission briefs. These training flights typically flew to Mountain Home, Idaho and back, passing overhead before landing.

At first, the special projects team found it difficult to locate and lock onto the plane traveling inbound at Mach 3. The radar operators searched for the return of the aircraft by scanning the horizon with the Nike radar used to track it on the inbound leg. Once the target appeared, the operator had only seconds to lock the radar onto the target. Heading inbound at Mach 3, the range decreased, and the elevation increased at too high a rate for the radar to slew and catch up.

The Nike radar operator acquired target acquisition by sweeping his beam at the expected azimuth for the target to pop over the horizon. With the radar azimuth and elevation locked on the target, the target's transponder answered with a spike for the operator to capture and lock in the range gate.

Once the Nike radar acquired a target lock, it drove the bull gear to align all other antennas on the target.

John Grace solved the target acquisition problem by bringing Barnes in from the NASA High Range at the Beatty Radar Station. Barnes, a hypersonic flight support specialist routinely locked onto the much faster X-15 and tracked it as it passed directly over his radar site at Beatty. (On the NASA, High Range in Nevada,

Barnes had maintained his proficiency with the radar system, the data transmission, communications, microwave, and telemetry systems. He had participated in tracking the Mach 3 A-12, YF-12, and M-21 Blackbirds, the Mach 3 XB-70, and the Mach 6.7 X-15.)

A typical airborne radar cross-section of the Mach 3 A-12 started with the A-12 headed away from Area 51. The Nike radar tracked the aircraft with the other radar systems involved passively ganged to the Nike servos. The objective was to obtain a basic radar cross-section under operational conditions. However, they never tested the electronic countermeasure (ECM) or electronic warfare systems equipment. The aircraft had a special paint treatment.

The flight plans called for the radar slant range versus elevation angle for the approach portion of the flight. For the refractive index model selected, the altitude of the aircraft appeared to be between 72,000 and 75,000 feet, about the radar test site, for the inbound run.

The Nike radar locked the Article in auto track outbound at nine nautical miles and tracked the plane to 255 miles where it became inbound. They plotted the ground range versus the azimuth range as the aircraft approached the radar station from 340 degrees.

The central reference servo system (CRS) plotted the rapid increase in elevation angle as the plane approached. The Bendix system was operated in the monostatic configuration to obtain cross-section data for horizontal polarization in S-band (21.73-2.98 GHz). A portion of the cross-section data was invalid because the maximum slew rate of the DSK was insufficient to provide antenna tracking as the vehicle passed nearly overhead.

The navy system was operated in the monostatic configuration to also obtain cross-section data for horizontal polarization at 173 MHz at a range of 140 nautical miles. The plot of cross-section data versus time for the Navy system indicated that the cross-section varied widely, with a maximum of 14 dBam. Again, the DSK was insufficient to permit antenna tracking as the vehicle passed overhead.

The MOD III system operated in the monostatic configuration to obtain cross-section data for vertical polarization in the c-band (5.47GHz). Like the other cross-section radars, the MOD lost track as the A-12 flew overhead. Nonetheless, it obtained values as high as 18 Db above 1 square meter for the cross-section of article 132 at C-band.

Once the A-12s deployed to Kadena for Operation BLACK SHIELD, the range shut down in June 1967 and remained so until early in 1968 when the MIG-21 arrived for exploitation. It was this experience learned while testing the U-2 and A-12 that qualified the special projects team for their role in exploiting the Soviet MiGs.

The CIA brought considerable knowledge and expertise to the table when the Soviet MiG planes arrived at Area 51 for exploitation. The methods and equipment used for the previous U-2, and A-12 projects and the technology gained applied to the MiG exploitation projects as well.

The RCS goals defined in the primary contract for the A-12 required two radar ranges, ground, and a flight range. The RCS team measured a U-2 on the ground range to verify the correlation between ranges and then flew and measured it on the flight range for comparison.

The ground range assisted in developing the A-12 configuration and defining its radar cross-section (RCS). The flight range verified the RCS of the A-12 in flight.

The ground range extended 1-mile long north on the southwest edge of Groom Lake. A major positive for using the lakebed, it provided a low noise background for the radars.

The 1-mile site supported an actual A-12 as well as models for measuring the RCS. A ½ mile site using an inflatable air bag to hold scale models and model sections of the A-12 to measure the RCS. A ¼ mile site consisting of a square meter sphere which erected from below ground to establish calibrations of the radars to a 1 square meter target. (0 Dbsm)

The radar antennas on the lake edge faced down range to the North. The heights of the support poles/airbag/cal sphere correlated with the radar antenna heights to maximize the radar signals. Special projects mounted the radar targets upside down and rotated them while the radars illuminated them. They recorded the return signals and referenced them to the calibration sphere.

The one-mile site contained a hydraulic pole buried in the lakebed. The pole, a couple of destroyer propeller shafts welded together, raised the targets up 52 feet. A 26-foot cube of concrete below the lakebed supported the pole. A hinged parabolic cover bolted to the top of the pole reduced the radar backscatter of the pole. The ground crew at Area 51 mounted the target with an internal rotator, mounted on the pole at ground level and then raised it. The ground crew then wrapped the cover around the pole to minimize the pole RCS.

The ½ mile site used an airbag to support the RCS targets scale models or A-12 section models. A 10-foot diameter, 20-foot high airbag mounted on a table below the lakebed and attached to a gearbox underneath to rotate and tilt the table. A little shed in the pit below the lakebed surface contained the power controls, air compressor, and a sump pump. They called it the swimming pool.

The ¼ mile site lay beneath the lakebed surface and consisted of a 1 square meter fiberglass sphere coated with aluminum tape attached to a fiberglass pole. A gearbox popped up the sphere with the pole tilted toward the radars to minimize having an impact on the sphere RCS and becoming the calibration reference for all RCS measurements. The special projects team's management brought in John Leonardi, a computer expert from National Cash Register to design and build a special purpose computer for radar timing control and data reduction and recording.

As an added feature of Leonardi's computer, he programmed it to recognize what to expect from the systems that it interfaced. The computer programming established standards for what it expected during a mission. If a preflight system evaluation failed to meet the desired signals and levels, it alerted those responsible of the anomaly in for correcting it or knowing it existed and ensured no loss of data simply because someone forgot to flip a switch.

The radars covered frequencies of 500–800 MHz, 1.2–1.5 GHz, and 2.7–3.0 GHz. Each radar contained two antennas pointing down range, one for transmitting and one for receiving. The antennas stacked in front of the radar building ranged from 4 feet above the ground to 30 feet. A dividing plenum of chicken wire erected between the transmitting and receiving antennas reduced transmit pulse into the receiver.

In early 1960, the range designers from SEI with EG&G support built and installed the radars. The SEI personnel consisted of 3 PhDs and a group of senior engineers. John Grace, a senior executive with EG&G, played a significant role in obtaining the various radar systems for Area 51. During this period, the ground range remained fully operational with a four-man group from Lockheed reviewing data and directing EG&G as to which radar systems to use. Additional Lockheed personnel transported and mounted the models at the 1 mile and ½ mile sites.

The RATSCAT, the ground radar range resembled the USAF's RATSCAT built at Holliman AFB on the White Sands lakebed using both SEI and EG&G personnel as consultants.

The chief scientist in the SEI contingent added much more to the projects than radar range designers. He became the countries guru in RCS reduction both for shaping and Radar Absorbing Material (RAM). He introduced RAM-loaded chines on the A-12 to attain an acceptable RCS.

The flight range evolved and articulated with the times, needs, and technology advances. The radars and associated tracking equipment typically used existing systems in the military inventory modified where necessary for the A-12 requirement.

The flight range extended the ground range with the added requirements of tracking the airborne target, measuring it out to 200 nautical miles while recording pitch, roll, yaw and transmit data to the site and calculating RCS returns versus look and depression angle. They measured RCS at 70 MHz for the Knife Rest radar, 172 MHz for the TALLINN (Tall King), and 3.0 GHz for the Fan Song.

The respective radar operators calibrated their radar before each mission. For target tracking and acquisition, the initial design used a 400 MHz telemetry transmitter on the A-12. It used a monopulse tracking receiver on the lip of the big dish to point the dish to a close enough approximation to the target for the 172 MHz or S-band radar to see it.

This higher accuracy track over the monopulse placed the Nike close for acquiring the target for the mission. The special projects team designed and built a bull gear (mechanical parallax correction) to allow switching of antenna control to the correct radar. The team used an X-Band Transponder on the A-12 for a high precision track. However, the 400 MHz monopulse tracking antenna receiver proved unworkable after extensive tests.

The special projects led by engineer Wayne Pendleton performed ground tests by loading an Army pickup truck with a 400 MHz transmitter and some gas-powered generators. They then drove the pickup truck to the hilltops to lock the dish to the transmitter. They never succeeded at grazing angles and concluded that the monopulse array would never work due to ground bounce interference.

For marking the target, the high-power output of the radars accomplished the target measurements. The S-band, a 60-foot dish Bendix SPS-29 radar from Lincoln Labs produced 1 megawatt and produced 750 kWh at 172 MHz. The 4-bay, directional antenna Yagi radar resembled a rooftop terrestrial television antenna or an amateur radio antenna provided a 100-kHz beam at 70 MHz.

The SPS-29, 172 MHz The United States Navy air search radar originally designed for small ships required mounting the output tube in an air cavity filled with a correct dielectric constant oil to lower the operating frequency to 172 MHz. At Area 51, running 3 1/8-inch rigid inch coax up the dish proved challenging as did repairing the broken folded 4-dipole antenna feed. The water-cooled radar transmitter at Area 51 for 425 MHz measurements of target bearing and range required the building of a shack with a swamp cooler to cool the cooling coils. The antenna resembled a bedspring and bore the nickname of bedspring radar.

The special projects team constructed its version of the Russian V-75 SA-2 GUIDELINE A-band, 75 KM Knife Rest radar. It took three tries to successfully build a radar that generated so much ozone the operators could not remain in the closed room during operations.

For measuring pitch, roll, yaw, and transmit data to the site, the telemetry package consisted of a compass and a gyro to generate the parameters. Its electronic converter put the signals into a binary code to modulate the transmitter with a 1 kHz signal phase modulated to produce a pulse for each bit with an error correction remainder code maintaining signal fidelity.

The special projects team originally intended to use an ARC-34 radio. They found it too complicated to operate and switched to using a 400 MHz taxicab radio. The team designed a low observable (LO) antenna about a square foot to match the planes Q-bay contour. Filled with Ecco sorb, it rested 4 inches deep in the camera Q-bay where it transmitted through a crossed slot aperture.

They used an F-101 chase plane for dry runs of the telemetry package. This was because the F-101 weapons bay used a big rotating door designed for launching massive nuclear bombs. It launched the bombs by rotating the door with a bomb attached and executing a pitch up maneuver to lob the bomb away from the F-101. Once he launched the bomb, the pilot rushed the plane out of the bomb kill zone. The special projects team bolted the package frame to the inside of the door and used a standard The United States Air Force UHF antenna attached to the outside of the door.

When installed in the U-2 at Burbank, the package tested OK on a UHF receiver using the UHF LO antenna. When the U-2 arrived at Area 51, the team failed to detect any 400 MHz signals and therefore could not calculate the angles to correlate where the team measured the radar cross-section. Such a miserable failure prompted a CTJ (Come to Jesus) meeting at Area 51 with severe gnashing of teeth. They disassembled the LO antenna and found it faulty, working as an attenuator instead of an antenna with decent gain and enough radiated energy for a ground check, however, with insufficient gain to achieve any distance. Burbank fixed the problem, and it worked when it arrived back at the test area.

For calculating radar cross-section returns versus look and depression angles, a special purpose computer provided timing for all radar systems. The computer, essentially a DECEMBER PDP-3 prototype recorded the test data on the site. Someone in special projects then hand carried the 1-inch tapes by plane to Waltham, Massachusetts for final calculations before subsequent tests, and for final data reduction.

The ground to slant converter was a synchro arrangement that acted as a mechanical convertor to calculate look and depression angles where the team measured the radar cross-section using pitch, roll, yaw with azimuth, elevation, and range of the tracking radar.

All the radar systems at Groom Lake used an actual square meter sphere for the ground range calibrations. They found this 42-inch diameter sphere too large to throw out of an airplane to measure the radar cross-section in free fall.

Instead, the radar cross-section team used 12-inch diameter world globes coated with aluminum tape for S-band calibrations. For calibrating the 172 MHz radar, they used a resonant dipole (6 feet long) made from a piece of pipe with end slots of fiberglass paddles. The paddles tilted to rotate the dipole as it descended to the ground. The 4X6-inch fiberglass "paddles" used vertical stripes of aluminum tape to provide a target for the Nike tracking radar. The dipole suspended below a parachute with a swivel allowing the dipole to rotate as it descended to the ground.

EG&G used the C-47, flying it as high as it could go without the participants suffering from lack of Oxygen, and threw these targets out the doorway removed for these missions. The Nike acquired the sphere or paddles and tracked them with the bull gear aligning all the antennas on the targets. The plane approached the drop lying about 5 NM out at around 330 degrees in a cross-range heading of 240 degrees. As the 172 MHz dipole rotated, the target returns peaked and fell with the computer calculating the radar cross-section from the peak returns.

As the United States Air Force presence increased, they took over the calibration drop missions using a helicopter for the drops.

They phased out the cardboard world globes for spun cast 12-inch Aluminum spheres and 26-inch spheres for the lower frequencies.

Having reams of data and its reduction proved extremely tough until someone came up with a novel solution to eliminate any clutter contamination. They incorporated an early range gate in the radar, and an operator continually monitored it during a test. When he observed clutter in the early gate (3 miles ahead of the target), he tagged the computer. When the early gate emptied, he tagged the computer again. The team discarded the tagged data for radar cross-section calculations.

The one wavelength in diameter of the inlets on the A-12 presented a primary source for reflections, causing a large radar cross-section at 172 MHz.

Even filling the spikes with RAM required additional effort to reduce their cross-section. Someone proposed generating a plasma cloud in front of each inlet to absorb/deflect radar paints produced by shooting electrons out to the inlets from X-ray guns having a hole in the anode to allow the electrons to exit the x-ray tube. Concerned about the X-rays having an impact on the pilots, the team determined the pilot did not need a lead shield in his seat for his protection from the radiation.

The apparatus filled the Q-bay, and it took a long time to fly an operational unit. The radar cross-section team built a pulsing timer for it to verify radar cross-section reduction. The team learned of the pulsing timer working when the operator of the 172 MHz radar came on the net during the test mission, voicing concern about his receiver oscillating and asking permission to troubleshoot it!

One of the parties involved in the exploitation developed a jammer and brought it to Area 51 to test its operation. It worked by getting in the back lobe of the Fan Song guidance signals where it spoofed the signal.

The simulator on site did not have the guidance antenna/system mounted on the Fan Song antenna pedestal. However, they did have one located approximately 200 feet east of the tracking antennas.

Whenever the A-12 came inbound from the east, the jammer failed to work. After some reflection, the special projects team repeated the test with the A-12 coming inbound from the north. The jammer worked! Apparently, the 200 feet (200 ns in time) caused a fault to the jammer, rendering it unable to recognize the threat. This sort of countermeasure experience proved invaluable for future projects such as the Soviet MiG exploitation project during radar cross-section evaluations.

A full model A-12 on the RCS Pylon constructed using two navy battleship driveshafts welded together

Full scale mock-up in full external configuration. The Willys M38 Jeep provides excellent scale. Note the shadow. Soviet infrared satellites picked up the shadow and knew about the plane in spite of it being removed from the pylon and hidden from the satellites.

The A-12 exhaust ejectors produced a large return during the EG&G radar cross-section tests. Lockheed physicist Ed Lovick studied ionizing the exhaust gases to shield the ejector from radar. The engineers from Pratt & Whitney and he experimented with adding potassium, sodium, and cesium to the fuel to develop a metallic salt that vaporized to produce a plasma cloud. The pole model has an added cylindrical frameworks covered with reflective material to mimic radar returns from the ionized exhaust plumes. The pylon and rotator are shielded tp prevent backscatter.

Radar Cross-Section of the A-12

In the spring of 1959, the Skunk Works team in Burbank built a full-scale mock-up of the A-12 for RCS tests. At the same time, Edgerton, Germeshausen & Grier (EG&G) of Las Vegas moved its radar test facility from Indian Springs, Nevada to a special pylon constructed on a paved loop on the west side of Groom Lake for the RCS tests. It had taken 18 months of testing and adjustment before the A-12 achieved a satisfactory RCS.

The low radar cross-section requirement for the follow-on reconnaissance aircraft, Lockheed continued anti-radar testing on the full-scale mock-up at the Watertown facility. The inlet, vertical tail, and the forward side of the engine nacelles of the airframe areas were giving the greatest radar return. Lockheed was looking at an improvement in the chine and wing regions, researching ferrites, high-temperature plastic structures, and high-temperature absorbing materials to find methods to reduce, the return. The Lockheed engineers proposed constructing the vertical tail section fins of laminated plastic. Lockheed subcontracted Narmco to build the fins. They expected the work in ferrites would help in reducing the reflectivity of the inlet and engine nacelle surfaces. Lockheed proposed a metal, and the plastic cover arrangement for the chine and wing edges, hoping the combination to accomplish a significant reduction in radar return. While Lockheed and its subcontractors came to grips with airframe fabrication, the engine manufacturer was experiencing problems of its own.

Both Lockheed and Convair submitted the final designs during the summer of 1959 that incorporated a Pratt & Whitney Mach 3.2 J58 engine. Pratt & Whitney made the first run in December 1957 as a Mach 3.0 design test stand engine, and it showed impressive development progress during the following two years. Pratt & Whitney presented cost estimates in the fall of 1959 totaling $80 million to take the effort through 31 December 1962 to further develop production of 36 engines, and for the overhaul, maintenance, and spare parts support.

Company representatives wanted to incorporate in the P-2 model the features necessary for Mach 3.2 flight. They could develop the 32,500 pounds thrust to take the J58 engine from its current stage of development of 26,000 pounds. Because of this confidence, they priced the engine at the $750,000 per unit figure that P&W used for other jet engine sales, which they estimated as being the value for the J58.

A major problem of a different nature achieved lower radar cross-section desired. The vertical stabilizers, the engine inlet, and the forward side of the engine nacelles became the airframe areas giving the greatest radar return. Lockheed researched ferrites, high-temperature absorbing materials, and high-temperature plastic structures to find methods to reduce the performance.

Eventually, they used a kind of laminated "plastic" material to construct the vertical tail section, a first time for using such a material for an important part of an aircraft's structure. Such changes in structural materials brought the designation change of the A-11 to the A-12. The CIA did not publicly disclose the A-12 as even existing until 1991, nearly 25 years after storing the surviving planes in a secure location at Palmdale, California.

While the CIA, Lockheed, and many other entities solved the design and unique component problems, the special projects team at Area 51 pioneered stealth. The CIA recruited its pilots to fly the plane with it now designed and ready for flight testing.

To test the effectiveness of antiradar devices, a small-scale model was inadequate; requiring only a full-size mockup. Lockheed accordingly built a full-size mockup that it delivered to Groom Lake in November 1959. A specially built truck transported the mockup over 200 miles from the Burbank plant to Area 51.

Here, the special projects team hoisted it to the top of a pylon and looked at it from various angles by radar. Tests and adjustments went on for a year and a half before the CIA deemed the results satisfactory.

During the process, it was found desirable to attach some sizable metallic constructions on each side of the fuselage. Kelly Johnson worried a good deal about the effect of these protuberances on his design.

In flight tests, however, they later developed their imparting a useful aerodynamic lift to the vehicle. Years later, Lockheed's design for a supersonic transport embodied similar structures.

On 29 November 1966, the EG&G special projects conducted an RCS test of Article 132. They conducted the flight test under operational conditions to obtain basic radar cross-section. However, they never tested the electronic countermeasure (ECM) or electronic warfare systems equipment. The aircraft had a special paint treatment.

The flight plans called for the radar slant range versus elevation angle for the approach portion of the flight. For the refractive index model selected, the altitude of the aircraft appeared to be between 72,000 and 75,000 feet, about the radar test site, for the inbound run.

The Nike radar locked the Article in auto track outbound at nine nautical miles and tracked the plane to 255 miles where it became inbound. They plotted the ground range versus the azimuth range as the aircraft approached the radar station from 340 degrees.

The Central Reference Servo System (CRS) plotted the rapid increase in elevation angle as the plane approached. The Bendix system was operated in the monostatic configuration to obtain cross-section data for horizontal polarization in S-band (21.73-2.98 GHz). A portion of the cross-section data was invalid because the maximum slew rate of the DSK was insufficient to provide antenna tracking as the vehicle passed nearly overhead.

The navy System was operated in the monostatic configuration to also obtain cross-section data for horizontal polarization at 173 MHz at a range of 140 nautical miles. The plot of cross-section data versus time for the navy System indicated that the cross-section varied widely, with a maximum of 14 dBam. Again, the DSK was insufficient to permit antenna tracking as the vehicle passed overhead.

The MOD III System operated in the monostatic configuration to obtain cross-section data for vertical polarization in the c-band (5.47GHz). Like the other cross-section radars, the MOD lost track as the A-12 flew overhead. Nonetheless, it obtained values as high as 18 Db above 1 square meter for the cross-section of article 132 at C-band.

Article 132 did not have the modification to permit installation of the telemetry package. For this reason, it did not include roll, pitch, and heading data. The Bendix System was receiving in the horizontal polarization mode; the Bendix normally receives in the vertical mode.

The Infamous Groom Lake House Six

The compartmentalization of Area 51 created a situation that prohibited separate groups commingling. Consequently, each group made their entertainment. The most famous was the pilots' hang out, House Six.

House Six started simply with the CIA somehow coming up with a new poker table and seven chairs. Up to then, the pilots, like everyone else, entertained themselves with card games in their small living room. It did not take but a day or two for the word to spread and from then on, the pilots played poker every night. They had a couple of ground rules, i.e., no high-stakes games and them playing the last hand at 2230 hours (10:30 pm) sharp each evening. They and the others groups al had a ritual whereby at 9:00 pm they made a mess hall run game. No one could drop out, and whoever won the game had to take orders for hamburgers, fries, malts, soft drinks or whatever, run to the mess hall and bring back the goodies, the calorie-laden snacks. Unlike the CIA and special projects groups, House Six allowed anyone to play, be they officers, enlisted men, contractors or whomever.

Once House Six acquired the card table, they wanted a bar set up on the other side of the house. Maj Sam Pizzo, head of the mission planning, but a navigator by trade, manned up to the challenge. He approached some Lockheed workers at the wood shop and asked if they could build a bar for House Six. In less than a week, House Six had a bar that Major Pizzo and Lockheed test pilot Lou Schalk concluded needed a fancy top rather than a plain piece of wood. A week later, the bar shined with an Epoxy finish.

Now the bar needed a refrigerator to keep beer and soft drinks. Ta-da, House Six soon had a refrigerator for their beverages. However, they still had to make frequent trips to the mess hall each evening for ice.

Maj. Harold "Burgie" Burgeson became the barkeep, and after making the initial purchase of the needed supplies, each weekend he replenished the inventory as needed, — everything paid for by people having a beer and dropping a quarter in the jar to pay for it.

Burgie, an avid golfer, was also responsible for creating the Area 51 golf course which consisted of one small desert sand (green grass does not grow so well in the desert without water), a hole in the ground with a small flag pole in it. The approaches to it came from any direction one might desire. The only hazard on the course was rattlesnakes and cactus. Thus, House Six became the unofficial Area 51 Base Club.

For their off-duty pleasures, each customer developed their entertainment. Project pilot Frank Murray and others took up model plane flying, and many Ham radio operators emerged from their tour at Area 51.

An exception to the commingling of groups came with the forming of the Area 51 softball team that competed with a softball team formed on the Atomic Testing Grounds.

The special projects and the CIA permanent party personnel mainly used Sam's Place and the CIA amenities, except for poker games conducted in individual houses, which typically offered four bedrooms, a shared living room, and kitchen.

The Broom Crew

Area 51 did not have mechanical runway sweepers in the early start-up days of the A-12. In the evenings, the desert had a nasty habit of rolling tiny rocks on the runway and unbeknown to anyone, and during the takeoff roll, those Pratt & Whitney Engines sucked up all little objects that happened to be on the runway. When sucked up into the engines, these rocks played havoc with the rotor blades, causing considerable damage which required engine changes. That, in turn, resulted in those engines being flown up to Westover AFB outside of Boston in Gil Saunders' C 130 aircraft, where Pratt & Whitney would meet the plane and take the engine in for repairs. Those little rocks resulted in an outlay of $250,000 per engine.

Consequent, to offset these costly disasters, the CIA decided to form the broom crew which consisted of a gaggle of men, officers included to line up across the runway and start sweeping the rocks from the runway to a point where it was estimated the A-12 would become airborne! This activity continued until the mechanical sweepers arrived on the scene.

Oyster Shells in the Desert? Intuitive at its Best!

The pickup and return of these damaged engines to the Area also played a part in boosting the morale. Boston was well known for having great seafood, and Major Pizzo thought perhaps he might take advantage of this when they sent their engines back east for repair. He called the Club Officer at the Wendover AFB and asked if he would not mind purchasing some seafood for Area 51 and if so, Pizzo would call him on the day before a flight, place an order and pay him for the goodies when the C-130 landed at his base the next day. This trusting soul agreed and, Pizzo started taking orders for, crabs, lobsters, oysters (still in the shells), scallops and shrimp. He called the club officer as soon as he had the "druthers list" completed, and sure enough when the C-130 landed at Westover AFB, the club officer delivered the goodies to the aircraft, Pizzo paid him, and from then on, those transactions worked without a hitch. The troops at Area 51 paid Pizzo the costs of the order. Ever since the early Area 51 veterans have enjoyed wonder how those who came later could explain the existence all those empty oyster shells found in the middle of the desert.

On one such flight, as the C-130 returned with the goodies, the aircraft developed engine problems and was forced to land at an air base in Colorado for repairs. The crew always placed the aquatic goodies in the rear of the C-130, the coldest part of the aircraft. However, on the ground in Colorado in the middle of summer, there was no cool spot in a plane for raw seafood. Those maintenance men repairing the aircraft could not understand why this C 130 crew was in such a panic stage trying to rush the repairs, which those in Colorado did in time for that type of problem. All turned out well, and no one became ill from the incident.

Logistics

The approved driving distance from Area-51 through the Nevada Test Site to "town" (Las Vegas) was 120 miles. The REECo personnel, mostly laborers, cooks, housekeeping, carpenters, and electricians drove in and out.

Base security allowed one Lockheed RCS tester to fly his Cessna to and from the Antelope Valley. In exchange, he assisted in security sweeps, even installing a failed U-2 window installed in the bottom of the plane to enhance ground viewing.

The CIA required all customers, most air force personnel, air force civilians, contractor maintenance personnel, and craft people to commute in and out on C-47s and Lockheed Constellations.

The CIA personnel drove to the site. EG&G special projects personnel flew to the site on CarCo, a subcontract airline of the Atomic Energy Commission.

The airplanes for EG&G started small and got larger as the project also grew. Initially, they flew to and from the site in a D-18, then a C-47, and then an F-27.

Each group bore incidences of emotional moments. A C-54 had crashed on Mt Charleston early in the U-2 program killing 14, all aboard. Those killed included an air force crew, key CIA, and Hycon employees.

Grouped even further the author and three others on the special projects team flew to and from Area 51 in a Beechcraft Queen Air flying out of a secure compound along the side of the McCarran Airport along Sunset Road. Though unconfirmed, some believe that Central Intelligence mainly chose the Queen Air group for their unique contributions, looking ahead to future projects rather than merely the current project.

The mess hall, open 24/7 remains to this day the major and longest living amenity at Area 51. Anyone ever serving at Area 51 lays claim to Area 51 having better food than any hotel or casino in Las Vegas. Having a world-class mess hall offset the severity of the regimen.

Tuesday and Thursday night dinners, famous for the steak baron of beef and lobster drew authorized support personnel from Burbank and all over who scheduled their trips to the area around those days.

The air force support element during OXCART arranged with the commissary officer at Wendover Air Base, Nevada to provide the mess hall with fresh oysters and lobster. Evidence of the great days of the CIA era remains today in the huge piles of oyster shells dumped on the Groom dry lake from the mess hall. Future archeologists will wonder how and when Area 51 connected to an ocean.

Circa 1968, installation of a translator finally brought television to Area 51. Instead of providing a form of relaxing entertainment, it invoked outbursts of disgust. The Area51 residents might be watching a great game of football, only to have the stronger signal of a bullfight out of Mexico override Channel 3 out of Las Vegas. The TV signal propagation changed with atmospheric conditions, causing skip conditions. Meanwhile, the poker games continued as the prime entertainment.

During Project OXCART, the Groom Lake evolved from a high desert surrounding a large dry lakebed surrounded by semi-barren mountain ranges within the Emigrant Valley to a fully functioning airbase. Consequently, project DOUGHNUT did not require installation or construction, leaving little for the permanent party personnel to do other than their specialty.

Project OXCART caught the spying eyes of the Soviet Union, bringing satellite overflights of what Area 51 personnel dubbed as ashcans. However, the arrival of the first MiG drastically increased the Soviet interest in the activities of Area 51, resulting in an increased frequency of satellites passing overhead. As with OXCART, the passing satellites required the ceasing of any electronic emissions and the hiding of any outdoor activities. That included rushing any planes caught on the airstrip or tarmac into a hangar or hoot-n-scoot shed existing for that purpose.

Besides severely disrupting schedules and activities, the downtime because of passing satellites created a toll on the CIA's highly skilled and highly motivated personnel on the special projects team. To most, the inability to energize any signal emitting electronic equipment made their assigned project unbearably dull. This boredom, however, unintentionally developed new technology from the special projects personnel using their knowledge, curiosity, and available means to experiment with something that later became a project at the facility.

Sadly, these achievements and most of the individual human accomplishments, mistakes (referred to as OS or "old shit!" moments), and humor have and may never become known.

The author earlier mentioned the MiG engine sucking in all the film badges and dosimeters. Another occurred during OXCART when the Lockheed engineers raised the nose of the first A-12 Blackbird, using a forklift, to evaluate fuel distribution. The fuel ran to the aft end of the plane, lifting the nose of the 105-foot plane into the rafters of the hangar. OS!

A long-running incident attributable to the special projects team is one referred to merely as "the goat incident."

Following the rain, the Groom dry lake bed always retained water for long periods. The lakebed did not absorb the water, so it stayed until it evaporated.

The goat, an all-terrain, six-wheel drive, an amphibious vehicle so named by the special projects team, hauled electronic test equipment and tools to the RCS pylon on the Groom lakebed. During a trip to the pylon, the vehicle somehow caught fire, becoming well advanced before the operator noticed it burning in the rear of the vehicle.

The fire destroyed the goat and its contents. For years after that, any time a piece of test equipment became lost or misplaced; the team immediately attributed it to the loss of the goat. Later, an inventory of all the equipment supposedly lost with the goat revealed losses to fill a small truck. (Most of the items declared as lost showed up in use by someone else at the time)

Groom Lake Mess Hall circa 1960

Housing during CIA Project Oxcart

Babbitt Housing at Area 51 late 1960s

Television reception was by a signal translator that rebroadcast the strongest signal. One moment, the strongest signal might be Las Vegas and the next a skip signal out of Madrid.

Area 51 Special Projects Technical Equipment.

Background:

The special projects personnel, activities, and equipment discussed herein occurred half a century ago and did not suggest the same existing today.

The CIA knew the Russian radars tracked Gary Powers et al. flying missions over Russia in the U-2. Altitude saved the U-2 from planes and missiles. The CIA also knew the U-2's vulnerable, which Powers proved.

The CIA needed a replacement and upgraded aerial vehicle if it intended to continue overflying Russia. The CIA wanted the replacement to have a low RCS. Kelly Johnson added speed and altitude to the replacement plane, the A-12.

All aircraft bought by the Government have requirements and specifications which require verification. A company called SEI (Scientific Engineering Inc.) claimed a method of doing that for RCS using a radar ground range and a radar flight range.

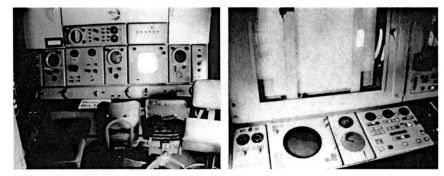

Nike X-Band

The CIA reversed engineered the Soviet-built TALLINN radar, and EG&G fabricated four Yagi antennas simulating the frequency of the TALLINN radar. The EG&G engineers named the radar systems Kathy, Kay, Mary, and Susan and used them at Area 51 to simulate the Soviet-style integrated air-defense system.

Some of the Special Projects equipment used to control tactical missions, receive, process, and store data from the missions.

The Groom Lake flight test facility rated as the best place to conduct RCS evaluations. However, SEI needed help to construct such a big project. To this end, EG&G employed a pool of Q cleared (AEC Top Secret) personnel, which helped the security aspects. EG&G provided some engineers and technicians from its nuclear activities next door at the Atomic Energy Commission atomic bomb testing.

CIA/Air Force Agreement on OXCART

On 14 October 1960, the CIA and the air force completed the final version of the joint CIA /Air Force agreement on organization and delineation of responsibilities for Project OXCART. In February 1961, Mr. Dulles and General White, Chief of Staff, Air Force signed the document.

In October 1960, the air force decided to buy a long-range interceptor version of the A-12. The Contracts Staff of DFD wrote the contracts, serving a twofold purpose by having the contracting performed outside normal military channels.

The CIA preserved the security of the A-12 program by keeping the A-12 purchase classified. Secondly, the order for three A-12 aircraft made easier a decision to reduce the 12 A-12 aircraft order to 10, wherein Lockheed was submitting cost estimates exceeding available funds.

A reduction in the numbers of A-12s procured was the only practical step that could keep the program budget in balance. Adding three A-12s to the production contract, therefore, lessened the impact on Lockheed of a two A-12 aircraft reduction, and thus eased the strains on the OXCART budget.

The CIA included in the air force A-12 procurement order for 10 J58 engines as a follow-on to the 30 engines previously ordered for the OXCART program. Extension of the engine order also gave some relief to the pressing financial problems of Pratt & Whitney. The total A-12 program (project KEDLOCK) was estimated to cost over 111 million, one-half, of which was for the Hughes Aircraft Company to design and build the fire control and missile system for the A-12. The contracts were the first of many the CIA would write for the air force in follow-on or associated programs to preserve the security of A-12 development and production.

Return to Area 51

Area 51 Designated

By 1961, as the program picked up momentum, the CIA widened the scope of its Development Projects Division, DPD, activities in the OXCART program. The CIA expedited the action to ready the operating site for occupancy by the aircraft and test personnel. However, there was a problem.

When the CIA first established the Area 51 site for testing the U-2 plane, it hid its participation by identifying the facility as Watertown operated by the National Advisory Committee for Aeronautics (NACA) for conducting weather research. NACA, a US federal agency, was founded on March 3, 1915, to undertake, promote, and institutionalize aeronautical research. On October 1, 1958, the NACA dissolved with its assets and personnel transferred to the newly created National Aeronautics and Space Administration (NASA). Now, the facility was reopening without NASA or the Atomic Energy Commission to provide a cover story. Consequently, as the CIA opened the site to test fly the A-12 OXCART, security considerations called for a new name for identifying the site to disassociate it from the U-2 program.

The Powers incident in May 1960 had associated the CIA and the U-2 with the Area 51 facility in the uncleared air force and industry circles. Consequently, the CIA ordered all earlier terms of reference to the Area 51 facility, such as "Watertown," "Groom Lake," or "The Ranch," eliminated from the vocabulary of OXCART cleared personnel to minimize possible speculation by unwitting persons. Henceforth it became simply "Area 51," to associate any activity with the official AEC designator for that portion of the Nevada Test Site and to disassociate Agency interest in the Area 51 facility.

The Operations and Materiel staffs of the Development Projects Division Tables of Organization drawn up by for initial manning of the operating detachment called for 70 personnel. The CIA manned the air force detachment, by making personnel levies on the air force for the various military specialties. The CIA and the air force worked out a phasing schedule for air force details, timing it to the estimated arrival times of the A-12 aircraft. The OX-CART staff at headquarters planned to have people identified by 1 May 1961 for necessary clearance and processing by 1 August. The CIA wanted everyone in place at Area 51 by October, and ready to begin training on 15 March 1962.

The air force modeled the U-2 program to carefully select the air force personnel before nomination for duty with the CIA. Because of close working relationships with the Strategic Air Command in the U-2 program, The CIA sent a request to the Strategic Air Command for guidance in the procurement of a detachment commander to which the Strategic Air Command responded by nominating Col Robert Holbury along with the principal members of his staff.

Norm Nelson

OXCART Pilot Selection

One pilot, called a driver by the CIA, operated the A-l2 with the responsibility of flying the aircraft, using sensor & EWS equipment and navigating to his destination. His training consisted of a ground school course at Lockheed Aircraft Co., followed by 21 missions in the A-12, for a total of 56 hours. This gave him an operational readiness status. His continuation training in the A-12 consisted of 18 sorties per quarter and included a minimum of seven aerial refueling. He accomplished his collateral training in an F-101 aircraft. ·He also received 148 hours of academic and field training annually.

The CIA and the air force jointly set the medical and physiological criteria, and General Don Flickinger coordinated it to process the chosen pilots, leaving it with the Development Projects Division operations staff to establish the operational requirements.

The pilots selected were those demonstrating outstanding proficiency and professional competence. Only jet fighter pilots with certain mandatory experience qualifications in the latest jet fighter aircraft qualified for selection. Only those emotionally stable, well-motivated, enthusiastic, and possessing good personal habits cut. The size of the A-12 cockpit dictated the size qualification, though the selection criteria only sought pilots between 25 and 40 years of age, under six feet tall and 175 pounds or less in weight. The CIA and the air force both carefully screened the pilot files for possible candidates to list. Background investigations and security assessments eliminated some, while physical examinations and psychological assessments eliminated others. All remaining nominees received an agency medical and psychological staff final review before final candidate selection.

Pre-evaluation processing eliminated all but sixteen potential pilot candidates to undergo intensive security and medical scrutiny by the CIA. Those surviving the CIA's scrutiny, tests, and interviews were then approached to take employment with the CIA on a very highly classified project involving a very advanced aircraft. In November 1961, the CIA received commitments from five of the groups. However, the small number recruited required the CIA undertake a second selection process. The CIA maintained the qualifying standards but revised administrative procedures to tighten up pre-evaluation processing and speed up pilot candidate identification.

The CIA and the air force collaborated to conduct pilot selection under the cover legend of establishing selection criteria for space crews. The CIA planned with the air force to effect appropriate transfers and assignments in a manner that covertly hid the real purpose of their training and lay the basis for their transition from military to civilian status. Pilot compensation and insurance arrangements were the same as that of the CIA U-2 pilots.

The CIA had placed one further step in place before bringing the pilot into the operational phase of the program. After the Powers incident, the director of Security had established an Overflight Panel to grant the CIA's final approval or disapproval of a pilot that might be a risk if captured. The Panel had representation from the Office of Security, Cl Staff, Medical Staff, TSD, OTR, and project headquarters. The Panel met periodically over the next two years until the first pilot complement was complete. Though it never surfaced during the security investigations at the time, there was one other thing that disqualified some of those interviewed for the major positions, especially the pilots, in the program, and that was the wife.

Security concerns were such that early in the OXCART program, air force personnel were not allowed to bring their families on the assignment. The outcry was such that the CIA relented and allowed dependents to accompany those chosen since they would live out of the state of Nevada. Nonetheless, the security personnel investigated the wife's background and disqualified the applicant if she had a drinking, gambling, or slept around, anything that could place her or the candidate in a vulnerable position for blackmail. Having a domineering wife could also disqualify the candidate. For example, one air force officer had passed everything to the point that the CIA interviewers offered him a job that they described as working on a classified project west of the Mississippi River. The pilot told the interviewers that he would need to check with his wife. The interviewers said that this was all right—for him to ask his wife. When the pilot returned, the interviewers were gone, and he never heard from them again.

Pilot Selection and Training

The pilot selection for the A-12 modeled after the earlier U-2 program at Area 51. Though the air force and the navy eventually flew the U-2, the CIA initially held majority control over the project, code-named AQUATONE. Despite SAC chief LeMay's early dismissal of the CL-282, the air force in 1955 sought to take over the project and put it under SAC until Eisenhower repeated his opposition to military personnel flying the aircraft. Nonetheless, the air force substantially participated in the project; Bissell described it as a "49 percent" partner. The air force agreed to select and train pilots and plot missions, while the CIA handled cameras and project security, process film, and arranged foreign bases.

Beyond not using American military personnel to fly the U-2, Eisenhower preferred to use non-US citizens. The world knows about the Taiwanese U-2 pilots called the Black Cats and the Brits who flew the U-2. The nationalities of other foreign pilots recruited remains classified.

The language barrier and a lack of appropriate flying experience proved problematic; by late 1955, the CIA and Air Force's 1170 SAS gave up on training foreign pilots because of their inability to learn the U-2.

The CIA required the air force pilots to resign their military commissions before joining the CIA as civilians; a process referred to as "sheep dipping." The CIA always identified the U-2 and A-12 pilots as "drivers," not pilots.

The program only recruited from fighter pilots with reserve air force commissions, as regular commissions complicated the resignation process.

The program offered high salaries, and the air force promised that pilots could return to their units at the same rank as other officers. The CIA held higher standards than the air force's once the latter began its U-2 flights. The CIA's program enjoyed a much lower accident rate. The CIA attributed this to its rejecting more candidates.

Test pilot Tony LeVier trained other Lockheed pilots to fly the U-2. He also trained the air force instructor pilots who in turn trained the CIA pilots "sheep-dipped" from the Air Force.

Both, the U-2 and A-12 programs started with the first of a kind aircraft. Thus, the pilots flew aircraft having no Dash 1 instruction manual or even a simulator. They flew by the seat of their pants, took notes, and wrote the books for future pilots to use for training.

Both the U-2 and the A-12 lacked a second seat for an instructor, so a pilot flew a solo flight with a radio his only means of receiving instructions. In both planes, the pilots adjusted to avoid what the pilots called the "coffin corner" as they learned the need to pay complete attention to flying when not using the autopilot.

The CIA recruited pilots of quite extraordinary competence for OXCART, not because of the unprecedented performance of the aircraft itself, but because of the qualities needed in men flying intelligence missions. For recruiting later U-2 pilots and pilots for the A-12, the CIA reversed its thought on the type individual to recruit.

Initially, the CIA wanted unmarried pilots, thinking them to have fewer family commitments and more apt to adapt to the lifestyle of a reconnaissance pilot. That might be true. However, the CIA quickly realized the same reason for the pilots remaining single applied to their suitability for flying the reconnaissance planes. Recruitment quickly switched to married men with preferably two children.

The A-12 OXCART/BLACK SHIELD Pilots

BGen Don Flickinger, of the air force, drew up the criteria for selection, with advice from Kelly Johnson and from CIA headquarters. The criteria called for pilots qualified in the latest high-performance fighters, emotionally stable, and well-motivated. The ages varied between 25 and 40 years and the size of the A-12 cockpit prescribed having pilots less than six feet tall and under 175 pounds in weight.

The air force screened its files for possible candidates and obtained a list of pilots. Psychological assessments, physical examinations, and refinement of criteria eliminated a good many. Pre-evaluation processing resulted in sixteen potential nominees.

This group underwent a further intensive security and medical scrutiny by the CIA. The CIA approached those remaining to take employment with the CIA on a highly classified project involving a very advanced aircraft. In November 1961, five of the group committed to the project. The small number recruited at this stage required the undertaking of a second search.

The final screening selected William L. Skliar, Kenneth S. Collins, Walter Ray, Lon Walter, Mele Vojvodich, Jr., Jack W. Weeks, Ronald "Jack" Layton, Dennis B. Sullivan, David P. Young, Francis J. Murray, and Russell Scott as pilots for the program.

Only air force fighter pilots made the selection to fly the A-12. Pilots from other military services applied, however, the air force refused to consider inter-service transfers.

The first pilots to join OXCART arrived at Area 51 in 1962. Everyone came qualified in supersonic fighter aircraft (a requirement). To fulfill their contract with the CIA, they resigned their Air Force Commission and became a civilian employee of the CIA. The CIA called this "sheep dipping." At the end of their contract with the CIA, the Air Force rescinded each of the resignations as though they never occurred, and promoted the officer. Each completed their careers in the Air Force. The CIA selected the following pilots to fly the A-12. Only five stayed with and survived the program to finish their distinguished air force careers.

Mele Vojvodich, Jr. (Dutch 30): A mostly Tactical Air Command pilot with combat experience in Korea during the 50s where he famously flew RF-86s north of the border into Manchuria. Vojvodich flew the first operational A-12 mission during Operation BLACK SHIELD. He retired from the air force a major general.

Ken Collins (Dutch 21): Collins, a former TAC Recce pilot, had an extensive experience in photo recon fighters, the RF-80, RF-84, RF-101. He earned the Silver Star in Korea during the Korean War. He survived the crash of an A-12 while on a training flight out of Area 51. Collins switched to the SR-71 program after reentering the air force. He retired from the air force, a colonel.

Walt Ray (Dutch 45): Ray had the same TAC Recce pilot background as Collins, same squadrons. Ray was a very experienced pilot, with a long military history. He had 3,354 hours of flight time, 358 hours in A-12s when he lost his life when his ejection seat malfunctioned following his ejecting from an A-12 on a training flight out of Area 51. Ray and his spouse, Diane was married only three months before he died.

Bill Skliar (Dutch 20): Test Pilot School Graduate, flew mostly supersonic day fighters. He came from the Test Squadron at Eglin AFB, Florida. Skliar was the first project to fly the A-12. In 1966, the air force revoked his resignation effective 1962, and he resumed his air force career as a lieutenant colonel flying the YF-12 and SR-71 at Edwards AFB, California. On 11 April 1969, Skliar and his RSO Major Noel Warner had to eject from a Blackbird that had blown out its wheels on takeoff. After retirement, Skliar continued to fly Formula One racing craft. On 18 August 1988, Skliar failed to survive his racing plane losing a wing and crashing nose-first into the desert about eight miles north of Stead Airport, Nevada.

Jack Weeks (Dutch 29): TAC background, mostly F-100. He came from Nellis to the program.

Dennis Sullivan (Dutch 23): Sullivan's background was primarily Air Defense Command (ADC) with a lot of F-106 Mach 2 experience. Sullivan joined the SR-71 program when he left Area 51 and reentered the air force. He retired from the air force as a brigadier general.

Jack Layton (Dutch 27 with the A-12) (Dutch 72 with the YF-12): Layton had extensive experience in supersonic airplanes, mostly the F-101 Voodoo, with the Air Defense Command's tactical nuclear-equipped fighters. (Colonel Layton bailed out of three burning aircraft during his air force career.) He retired from the air force, a colonel.

Dave Young: Young came from the Test Pilot School at Edwards, AFB. He flew the A-12 in training but left the program early.

Russ Scott: Scott also came from Edwards Test Pilot. He flew the A-12 in training but left the program before it went operational.

Lon Walter: Walter came to the program with a fighter pilot background. He left the program early and retired from the air force a major general.

Frank Murray (Dutch 20, Skliar's number before leaving the program): Murray came to the program on the support staff of the 1129th SAS flying chase in the F-101. The CIA selected him to fly the A-12 to replace Walt Ray and other pilots lost through various attrition. Murray was the last pilot to fly the A-12. He retired from the air force a lieutenant colonel.

By the time the A-12 became operational, only six, Vojvodich, Collins, Layton, Sullivan, Weeks, and Murray remained in the pilot force. These CIA determined this six enough to man the one Detachment deployed to Okinawa during Operation BLACK SHIELD, keeping the pilots busy moving to the detachment for six weeks, then home to Area 51 for a while, and back.

After selecting the air force pilots, CIA sheep-dipped them from the air force. Each resigned their air force commissions and entered a contract with the CIA. During operational flights, the CIA paid the pilots $1,500.00 a month. They contracted with compensation and insurance arrangements like those for the U-2 pilots.

The CIA and Lockheed decided in the early stages of the program where to base and test the aircraft. Lockheed clearly could not do this in Burbank, where they built the plane, if for no other reason too short a runway at Burbank.

They considered Groom Lake the ideal venue where Lockheed flight tested the U-2. Though deficient in personnel accommodations, POL storage, and an inadequate runway for the A-12, Groom Lake offered excellent security and needed only a moderate construction program to provide sufficient additional facilities.

Mele Vojvodich

Frank Murray

Jack Layton

Ken Collins

Dennis Sullivan

Dennis Sullivan

Escape and Evasion Training

In July 1963, the OSA deputy for Field Activities in a meeting with the Commander of Area 51 made it mandatory that the Director, Field Activities Division visit Area 51 on a minimum basis of once every sixty days. He also set in motion an escape and evasion training program in three categories for the project pilots. CIA headquarters would train them for resistance to interrogation. The Area 51 commander would be responsible for survival training and escape and evasion training. The CIA was violently opposed to spending $4,000 for a special vehicle to transport project pilots during this training. He expressed an intent to require within the coming year an A-12 qualified operations officer to become a member of the headquarters staff. He discussed with the commander a requirement for providing a pilot ground school for Operations personnel at headquarters. The CIA would provide two or three pilots for this purpose.

Chapter 6 - The Build-up

The Area 51 physical buildup in 1961 saw the completion of the 8,500-foot concrete runway. To cut costs, the CIA had constructed a 100-foot wide runway rather than 200 feet typically used for military runways. The CIA obtained three surplus Navy hangars from the US Navy at Hawthorne, Nevada that it had dismantled and hauled to Area 51 for erection on the north side of the Area 51 facility. The CIA also located some surplus Babbitt housing located on an old Navy installation and transported slightly over 100 of these buildings to Area 51 where it readied them for occupancy. The contractors completed 18-miles of road paving in 1961.

Harry Martin

Commercial power was not available. Therefore, the CIA brought in additional generators to provide the electrical power required by the growing installation. TSgt Harry Martin and others assigned to the Air Force's 1129th Special Activities Squadron arrived to begin construction of the fuel tank farm in 1961 and completed it in early 1962 with a capacity of 1,320,000 gallons of the special fuel required for the A-12. For warehousing and shop space, the CIA constructed new buildings and rehabilitated the older buildings, having all necessary facilities ready in time, for the forecasted delivery date of Aircraft No. 1 in August 1961.

Lockheed estimated everything required in such respects as monthly fuel consumption, hangars and shop space, housing for personnel, and runway specifications.

Headquarters compiled a list of the main requirements to come up with a construction and engineering plan. Preparing facilities for radar studies became the cover story for anyone curious about the activity in this remote spot. The personnel belonged to an engineering firm with support from the air force. The remote location reduced the effect of electronic interference from outside sources.

Though having excellent security, the site at first afforded few of the necessities and none of the amenities of life. Lockheed provided a C-47 shuttle service to its plant at Burbank, and a chartered D-18 (Lodestar) furnished transportation to Las Vegas because of the long distance between Area 51 and any metropolitan center.

Daily commuting was out of the question. The CIA billeted the construction workers arriving during the 1960s in surplus trailers. A new water well and a few recreational facilities helped. However, it took some time before accommodations became agreeable for the weeklong stays.

Nevada law became one of the minor snags during the Area 51 development. Nevada law required reporting the names of all contractor personnel staying in the state for more than 48 hours. The CIA realized that listing all these names and identifying companies involved exposing the real purpose of the activities. The CIA's General Counsel discovered a clause in Nevada law that exempted Government employees from these reporting requirements. Thenceforth all contractor personnel going to the site received appointments as Government consultants. Only government employees work at the site became the reply to any questions.

Construction began in earnest in September 1960 and continued a double-shift schedule until mid-1964. One of the most urgent tasks centered on building the 8,500-foot runway required for the A-12.

The existing 5,000-foot long asphalt runway could not support the weight of the A-12. Between 7 September and 15 November, they poured over 25,000 yards of concrete to build another runway. Providing some 500,000 gallons of PF-1 aircraft fuel per month posed another major problem. Neither storage facilities nor means of transporting fuel existed.

The CIA considered an airlift, pipeline, and truck transport, deciding truck transport the most economical. The CIA made this feasible by resurfacing approximately eighteen miles of highway leading into the base.

The CIA obtained, dismantled, and erected three surplus Navy hangars on the north side of the base. They designated them as Hangars 4, 5, and 6. They built a 7th hangar and transported over 100 surplus Navy Babbitt duplex housing buildings to the base and readied them for occupancy. The air force finished the fuel tank farm in 1962 with a capacity of 1,320,000 gallons. Warehousing and shop space began, and repairs made to older buildings. All this, together with the many other facilities provided took a long time to complete.

Base Facilities,

The OXCART aircraft program based at Area 51, a restricted area in the Nevada Test Site, the necessary facilities and staffing to support the test, training operations and operational deployment of the A-12. The population averaged 1,500 persons, including military and CIA civilian employees, on station to support the OXCART and TAGBOARD projects. About 650 of these were in direct support of launching operations, and approximately 611 were involved in indirect support such as planning, firefighting, and guards. Most of these people were under contract to Lockheed Aircraft Company or its subcontractors and were on permanent duty in this area. The military personnel and CIA civilian employees were on a basic three-year tour.

The CIA invested a total of $21 million in Area 51 for runways, buildings, housing, navigational aids, water supply to make the base self-sufficient. CIA personnel supervised the base support and maintenance. Reynolds Engineering and Electrical Company, a contracting company from Las Vegas, 239 persons engaged in base maintenance work. Total cost per year for salaries and necessary equipment was $5.5M.

The original U-2 hangars now served as maintenance and Machine shops. Facilities in the main cantonment area included workshops and buildings for storage and administration, a commissary, control tower, fire station, and housing. The airspace over Groom Lake became part of a new Restricted Area called R-4808N, replacing the former Prohibited Area P-275 that covered both the Nevada Test Site and Area 51 and prohibited overflights below 60,000 feet.

September 1961 CIA Inspector General Lyman B. Kirkpatrick arrived at Area 51 for a three-day visit. Afterward, he made some critical comments regarding Area 51 security and OXCART project management.

In his preliminary summary report, Kirkpatrick stated, "The 'Area' appears extremely vulnerable in its present security provisions against unauthorized observation. The high and rugged northeast perimeter of the immediate operating area, which I visited to see for myself, is not under government ownership. It is subject to a score or more of mineral claims, at least one of which is visited periodically by its owner. Several claims are sites of unoccupied buildings or cellars which together with the terrain afford excellent opportunities for successful penetration by a skilled and determined opposition."

Kirkpatrick felt Area 51 already demonstrably vulnerable to air violation including landings. He took issue with major installations not rigorously protected against sabotage. He felt they undertook to construct the facilities before the construction personnel receiving a full security clearance.

Richard M. Bissell thought these points valid. Mainly, he referred to the operators of the Groom Mine overlooking Groom Lake.

Meanwhile, however, they readied the necessary facilities in time for the forecast delivery date of aircraft No. 1, Article 121 in August 1961.

The facilities became ready before the plane. Originally promised for delivery at the end of May 1961, the date first slipped to August, largely because of Lockheed's difficulties in procuring and fabricating titanium. Moreover, Pratt & Whitney found unexpectedly great trouble in bringing the J58 engine up to OXCART requirements. In March 1961, Kelly Johnson notified headquarters:

"Schedules are in jeopardy on two fronts. One is the assembly of the wing, and the other is in the satisfactory development of the engine. Our evaluation shows that each of these programs is from three to four months behind the current schedule."

To this Bissell replied: "I have learned of your expected additional delay in first flight from 30 August to 1 December 1961. This news is extremely shocking on top of our previous slippage from May to August and my understanding as of our meeting 19 December that you essentially overcame the titanium extrusion problems. I trust this is the last of such disappointments short of a severe earthquake in Burbank."

CIA headquarters realized delays causing the cost of the program to soar. The CIA placed Norm Nelson, a top-level aeronautical engineer in residence at Lockheed to monitor the program and submit progress reports.

Kelly Johnson resented this oversight and refused to provide Nelson with an office. He finally gave in and provided him a desk in a 10-square-foot closet in a hangar, however, sticking to his security restrictions, denied him having a telephone. For years, Norm Nelson used an outside pay phone to report back to Langley. The CIA representative operated out of this closet until 1965 when Dick Sampson replaced Norm Nelson as the CIA security officer. Norm Nelson retired from the CIA and went to work for Lockheed at the Skunk Works.

Delays nevertheless persisted. On 11 September, Pratt and Whitney informed Lockheed of their continuing difficulties with the J58 engine weight, delivery, and performance. The completion date for Aircraft No. 1 by now had slipped to 22 December 1961, and the first flight to 27 February 1962. Even on this last date, the J58 unready for flight. The CIA and Kelly Johnson, therefore, decided to use for the early flights the Pratt and Whitney J-75 engine, designed for the F-105 and flown in the U-2. Lockheed proceeded to fit the engine, along with other components to the A-12 airframe to power the aircraft safely to altitudes up to 50,000 feet and at speeds up to Mach 1.6.

With this decision, final preparations began for the testing phase. On 15 November 1961, the air force named Air Force Col Robert J. Holbury commander of the secret base, with the CIA's Werner Weiss, a.k.a. The Desert Fox, as his deputy. The base operated as a CIA facility for another 18 years.

Colonel Holbury, while a captain, earned a commendation by General Patton, Commanding General of the Third Army with an endorsement to the commendation by Lieutenant General Vandenberg, Commanding General of the Ninth Air Force. The award was for his heroic conducting of a historic low altitude dicing mission along the Saar River during bad weather on 5 January 1945 while assigned to the 10th Photo Reconnaissance Group.

Born in Avon Lake, Ohio in 1920, General Holbury flew 76 combat missions in the P-38 Lightning and P-51 Mustang for a total of 142 combat hours. In Vietnam, General Holbury flew 11 different types of aircraft on 149 combat missions for a total of 365 flying hours. Holbury retired as a brigadier general and a command pilot.

Fuel Storage.

The program installed storage facilities for the PF-1 fuel used by the A-12 at established or selected points in the hours ZI and overseas. SGT Harry Martin was among the first Air Force personnel to arrive at Area 51 to prepare for the flight testing of the CIA's A-12 surveillance plane.

He was responsible for installing the fuel tank farm required to provide 500,000 gallons of the special, high flash point, high thermal stability jet propellant (JP7) fuel required each month.

At speeds, more than Mach 3+ very high skin temperatures were generated due to air friction. Both the earlier U-2 aircraft and now the Blackbirds required aa low volatility jet fuel not affected by the heat.

Shell Oil Company developed JP-7 jet fuel for this purpose. Manufacturing several hundred thousand gallons of the new fuel required the petroleum byproducts Shell normally used to make its Flit insecticide.

Meeting the CIA's fuel need caused a nationwide shortage of that product that year.

One could not ignite the fuel with a match. The very low volatility and the relative unwillingness of JP-7 to ignite required triethylborane (TEB) to be injected into the engine to initiate combustion, and allow afterburner operation in flight.

Adding cesium-containing compound known as A-50 aided in disguising the radar and infrared signatures of the exhaust plume.

Support Aircraft.

In April 1962, support aircraft began arriving at Area 51. These included: six McDonnell F-101B and two F-101F Voodoos for training and photo chase. Two T-33A Shooting Stars arrived for proficiency training, one Lockheed C-130 Hercules for cargo transport, one U-8A for administrative use, one Cessna 180 for liaison use. The Cessna 210 later replaced the Cessna 180. The aerial assets included a Kaman HH-43 helicopter for search and rescue, subsequently replaced with a UH-1. Two F-104A/G Starfighters (56–0790 and 56–0801) served as chase planes during the A-12 flight test program.

The OXCART program used F-101 support aircraft for pilot proficiency training and chase for the A-12. A C-130 provided for personnel movement and classified cargo, and two T-33s provided rapid transportation and et qualification of the pilots. Area 51 also had two helicopters, a U-3B for emergency air evacuation, search, and security patrol of the area, and an H-43B for search and rescue and paramedic jump training,

In a memorandum for the deputy for technology, OSA, dated 17 August 1963, the OSA deputy for field activities recommend deleting from inventory the F-104 chase aircraft become it was uneconomical. Since the CIA had picked up the responsibility for providing the chase planes for the A-12 flight tests, he recommended an additional two F-101 aircraft.

The Air Force Working For the CIA

In a memorandum dated 14 October 1960, and approved by Thomas D. White for the air force and Director Allen W. Dulles for the CIA, the CIA outlined the organization and delineation of responsibilities for Project OXCART. The air force approved the proposed arrangement on 15 February 1961, and the CIA did so the following day.

The memorandum stated the director of Central Intelligence and the chief of staff, air force jointly exercising the general direction and control of the project, subject to guidance from higher authority and coordination with other departments of the Government as appropriate. They would furnish guidelines to lower echelons, ensure the conformity of operations under the project with national policy, and make recommendations to a higher authority on matters transcending their authority. Further, it was their joint responsibility to resolve differences that might arise at lower staff and operating levels.

The memorandum outlined the responsibility for conducting the project, starting with the organizational elements of the CIA. It recognized the CIA had in existence a project headquarters headed by a CIA project director, and an air force assigned air force officer as deputy project director. The CIA project headquarters would establish an operational unit in the Zone of Interior, manned by the air force and CIA personnel in numbers, proportions, and skills as agreed between the project director and the air force project officer.

The CIA would carry on CIA rolls as chargeable to CIA for a planned minimum of three years all military personnel assigned to Central Intelligence for full-time duty on this project.

The memorandum provided supervisory responsibility to the air force deputy chief of staff, operations of designating an air force project officer who, under the guidance and direction of the deputy chief of staff, operations, was the action officer and point of contact for all functions related to air force interests in the project. The air force would establish a project staff in addition to the DCS/O project officer. This Staff included selected officers designated by other interested air force staff agencies who would act as points of contact for the project officer within their several offices.

The CIA project director and the air force project officer's primary responsibility was for the development and execution of all activities concerning the project within their organizations the resolution of differences that might arise at lower echelons and the reporting of progress and the making of recommendations to their respective chiefs.

The project headquarters was responsible for any continued research and development, operational planning, and the direction and control of activities in the final operational phase of the project during overflight launches.

The air force project staff was responsible for implementing plans approved by the CIA project director and the air force project officer and arranging for air force support of project activities which could appropriately be furnished through staff channels or by other air force commands

The air force project officer held the responsibility for the security of this project, where he approved in advance and monitored all clearances for personnel within the Department of Defense, DOD.

Activities under this project fell into three phases that overlapped one another in time, but with the kinds of activities involved distinguishing each of them.

The following were the specific authorities and responsibilities of the several organizational elements in the successive phases of the project:

The major activities of the first phase were research and development, procurement, the construction and activation of a test and training base, the testing of equipment, and operational planning.

Delivery Schedule Delays

Lockheed reported in January 1961 solution to all tooling problems but painted a dismal picture as to the material delivery situation. Titanium wing extrusions were far behind delivery schedule. The lack of needed supplies made it impossible for Lockheed to work to a programmed manpower level. At one point, due to a material shortage in March 1961, only 20% of the labor force was working on the fabrication.

In one of his progress reports, Kelly Johnson informed Bissell that schedules were in jeopardy on two fronts. One was in the assembly of the wing, and the other in the satisfactory development of the engine. Lockheed's evaluation showed each of these programs as 3 to 4 months behind schedule. While every effort was being expended to make up much of this time, Johnson said that he would be greatly amiss not to state these facts to avoid paying excessive overtime at a significant cost in such areas as development of facilities at the test area. Johnson said to Bissell that there might be other important vendors in trouble trying to make the initial schedule.

Mr. Bissell responded with a sharp message wherein he expressed his shock of learning of the additional delay slipping the first flight from 30 August to 1 December 1961 following a previous slippage from May to August. Bissell had understood from a meeting on 19 December that Lockheed had overcome the titanium extrusion problems.

Bissell further questioned the desirability of continuing the reconnaissance version of this aircraft given the overweight challenges and the effects of performance uncertainties in the A. R field, and the extreme difficulties being encountered by other contractors. Bissell chastised Johnson, saying, *"I trust this is the last of such disappointments short of a severe earthquake in Burbank."*

Bissell made it clear to Johnson that he expected Lockheed to work around the clock to complete the wing assembly and to employ all reasonable shortcuts to expedite completion of the first several aircraft and recovery of delivery schedules ASAP. He wanted to know Johnson' measures taken to prevent skyrocketing costs due to this delay both in Burbank and other suppliers. Bissell expressed concern about the build-up at Area 51, personnel acquisition, over time, and some other issues. Bissell let Johnson know that he was anxious also to explore the impact of the highly accelerated flight test program of this additional delay.

The delay had apparently become an item of significant expense that Bissell needed to re-plan on the less costly basis.

Project headquarters found it imperative at this point to improve its capability to monitor airframe development against the fiscal and timing requirements. Thus, Bissell and others on the OXCART team employed Norm Nelson, a top-level aeronautical engineer, as the CIA's resident at the contractor's facilities to monitor program progress and submit technical reports. Arriving in May 1961, Nelson began providing independent reports, thus keeping the technical program staff at headquarters abreast of developments at Lockheed on a timelier basis.

Seeking to Improve the Delivery of Titanium,

Headquarters sent a team to visit Titanium Metals Corporation (TMC) President directly concerning the high national objectives and priority of the OXCART program. The company most commonly referred to as TIMET, a shortened version of "TItanium METals," promised fuller cooperation. By May, Lockheed had sufficient sheet metal on hand for capacity production. However, extrusions continued to give it trouble. Lockheed now had the first flight of the first plane, Article 121, set for 5 December 1961 with expressed concern as to the status of engine development and possible late deliveries.

Obtaining titanium from the Soviet Union to build a plane to overfly the USSR

The airplane is 92% titanium inside and out. The CIA quickly realized that it was building the airplane for which it didn't have the ore supplies – an ore called rutile ore. It's a very sandy soil and it's only found in very few parts of the world. The major supplier of the ore was the USSR. Working through Third World countries and bogus operations, they were able to get the rutile ore shipped to the United States to build the A-12 and the family of Blackbirds that followed.

By mid-year, fabrication of the first airframe was in hand. Pratt & Whitney informed Lockheed on 11 September, of their continuing weight, delivery, and performance difficulties with the J58 engine. The delay slipped the completion date for the first aircraft to 22 December 1961, and the first flight to 27 February 1962. The CIA and Lockheed would need a substitute engine to meet even this last date since the J58 would not be ready.

Lockheed proposed using a P&W J75 F-105-designed engine that the CIA had earlier flown in the U-2. Lockheed could fit the engine, along with other components, to the A-12 airframe. The J75 could power the aircraft to altitudes up to 50,000 feet and at speeds up to Mach 1.6.

Adding to the anguish at project headquarters was the delivery slips causing an accompanying increase in cost. The total proposed target price had risen to $165 million for 10 A-12 aircraft when submitted by the contractor on 13 November 1961. The J58 engine development problems of Pratt & Whitney persisted on into 1961 and beyond, with slow progress resolving them.

Project headquarters became even more concerned at the spiraling costs attending the engine development program because of Pratt & Whitney estimating its original cost estimates on the low side. The Navy and the Air Force entered negotiation in January 1961 for direct funding of the J58 development program. The negotiation, in effect, reduced requirements for project headquarters' funding, but not for long. In May 1961, the costs increased with the production cost of the 30 engines now estimated at over 45 million with projected cost by September 1961, to over $51 million.

The Air Force accepted a greater financial responsibility in engine procurement because of its involvement in the A-12 program, from which it derived many technological benefits in other air force programs.

The Navy, despite the problems and increased cost, continued funding the J58 engine. It remained the only appropriately sized propulsion system for the A-12 family of aircraft. It was much to the interest of the Government that the CIA did not hold Lockheed to the terms of the contract, thus preventing the company being in serious financial jeopardy.

The additional ten engines required for the A-12 program afforded the opportunity to renegotiate the existing engine procurement contract. The parties amended the agreement to permit repricing the 40 engines at a higher unit level. The revised ceiling price guaranteed ultimate recovery to Pratt & Whitney of all the projected overrun. The net result of this was Pratt & Whitney never ran the risk of defaulting on the delivery of engines. Pratt & Whitney would deliver the 40 engines without a penny profit.

This was not to say that project headquarters was gentle in its relations with the engine contractor during these trying days. In one instance, project headquarters strongly recommended to Pratt & Whitney that it reorganize its management.

Pratt & Whitney's stagnate production came about in part because of its facilities being in East Hartford, Connecticut, and its research and development facility in West Palm Beach, Florida. The company was accumulating meaningful engine test time. Mr. Bissell, as part of his strategy, recommended that Pratt & Whitney co-locate its primary development work with production in Connecticut. In a letter dated 29 November 1961 to Pratt & Whitney's General Manager, Mr. L. C. Mallet, he reminded the executive of P&W's obligation to examine and reexamine all avenues leading to the expeditious and economical realization of its goal. He further advised Mr. Mallet that the CIA would continuously reappraise the contractor's progress and performance achieved by competitive programs.

Pratt & Whitney responded by shifting its managerial personnel to and within the Florida facility rather than relocating to Hartford. When a definite improvement appeared in the engine development management picture, the CIA relaxed the tactic of resettlement to Hartford.

Headquarters' continuing to exert pressure to get the program moving caused the implementation of two other contractor reorganizations. Two companies, Hamilton Standard Division (fuel controls) and Vickers Inc. (hydraulic pumps), both subcontractors to Pratt & Whitney made substantial increases in engineering workforce and test facilities availability in late 1961 and in January 1962 respectively. December 1962 saw these efforts making significant improvements in fuel control, hydraulic pump, and engine turbine durability. Each had increased to fifty hours the point of acceptable operation at their extreme temperature environments. By integrating these improved components into the engine, it satisfactorily completed its fifty-hour preliminary flight rating test in January 1963.

Detachment 1, 1129th (Air Force) Special Activities Squadron.

While the CIA awaited the Lockheed flight test phase beginning in early 1962, the operational detachment took form after headquarters; Strategic Air Command selected Air Force Col Robert J. Holbury, as the Area 51 commander. In November 1961, he and his, staff received several days of briefings at the CIA's project headquarters.

The CIA Instructions to the Area 51 Air Force Commander

On 25 November 1961, Richard Bissell sent Colonel Robert J. Holbury instructions for Holbury's assumption of the duty of commanding Area 51. He instructed Holbury to arrive on 15 November 1961 to assume the duties as Chief of Base, representing the CIA at Watertown, Nevada, the Commanding Officer, of Detachment 1, 1129th (Air Force) Special Activities Squadron.

As he did with Colonel William, Yancey, the Area 51 commander for the earlier U-2 program, Bissell made Holbury's command position known to those US officials and industry personnel whom Holbury would solicit cooperation in furtherance of his mission. He further instructed Holbury his mission of developing, as soon as possible and maintaining an operationally ready unit capable of executing missions as directed by the CIA project headquarters.

Holbury's operational duties and responsibilities as the chief of all CIA operations gave him authority over all CIA staff, and detailed personnel and responsible for the supervision of any CIA directed activities phased through Area 51. He would utilize personnel, materiel, facilities, and funds to ensure the most efficient use of these assets toward the accomplishment of the overall mission within the framework of the program under the directives received from headquarters.

Bissell gave Holbury the responsibility of maintaining compartmentation of activities and personnel to the extent required by security and as necessary to preserve the long-term effectiveness of the program. Holbury had to consolidate administrative and support facilities to the extent, and in the manner, he deemed best to carry out his mission. This included his developing and maintaining up to date emergency plans for Area 51 and remained familiar with applicable war plans as developed by headquarters. To separate the air force supporting Project OXCART at Area 51, Bissell established a chain of command that placed the chief of base responsible to the Chief, Development Projects Division, the deputy director (Plans), and the director of Central Intelligence.

Bissell gave Holbury the authority and responsibility of all financial and budgetary matters, plans, estimates and expenditures relating to the Area 51 operations. He authorized Holbury to expend funds the responsibility of ensuring that these authorizations stayed within the authorized amounts and that the operation observed all policies and procedures in appropriate CIA regulatory issuances. He utilized the established communications system and procedures between Area 51 and headquarters, ensuring that all communications, both cable and dispatch, to and from Area 51 were accessible to him to the extent and in the manner, he desired. Holbury referred any matter of particularly sensitive nature to the chief, DPD, on an eyes-only basis, or to the deputy director (Plans).

The maintenance of physical and operational security lay with the commander by CIA security directives as applied to Area 51. He reported any unusual problems to headquarters and maintained, as prescribed in report guidelines, records of activity currently in progress, status to date and projected adherence to scheduled completion dates. He informed project headquarters of any delays and amendments to planned activities as they ascertained discrepancies.

A CIA staff employee and the CIA's previous chief of the Area 51 facility, Mr. Werner Weiss, aka the Desert Fox, continued at the site as deputy commander for Support.

Key staff personnel first attended A-12 ground school at Burbank and F-101 training. The F-101 most closely approximated the A-12 in flying characteristics and was equipped with two, after burning engines. As unit training developed, the F-101 was to serve primarily as a trainer for the A-12 pilots and secondly as a chase aircraft for monitoring A-12 takeoffs, refueling, and landings.

Headquarters' plan for detachment training called for the air force supplying several types of support aircraft. These included eight F-101s for training, two T-33s for proficiency flying, a C-130 for cargo transport a U 3A for administrative purposes, a helicopter for search and rescue and a Cessna 180 for liaison use. Also, Lockheed provided an F-104 to act as chase aircraft during the A-12 flight test period codenamed, Project OXCART.

During the spring of 1962, the support aircraft began arriving, requiring additional personnel for maintenance of the support aircraft at Area 51, reaching 144 positions in July. The CIA referred to the staffing complement at Area 51 as Station D within the Development Projects Division to distinguish it from the U-2 detachments.

CIA's Werner Weiss, the Desert Fox

Werner H. Weiss, 1917–1997, a GS-15 with the CIA, was singularly the most important individual concerning the mission of the 1129th Special Activities Squadron at Area 51.

Werner Weiss came to America when he was nine years old. He came with his parents and his two brothers, a worker's family, and settled in Brooklyn, NY.

He did not know a word of English. He enrolled in a public high school where his talent in mathematics saved him from the school placing him several grades lower than his age warranted.

He learned English rapidly and left home after he finished high school with the firm intention of serving his new country. He became an American citizen by way of his father's naturalization. His memories of post-WWI and Nazi Germany instilled in him a fervent appreciation of what it meant to live in this land of hope and freedom. This intense patriotism directed the rest of his life.

He started by going to Washington, DC, where he got a government job that he always swore his having a lower rank than a GS-1. He passed the permanent Civil Service exams, worked while taking courses at the University of Maryland, got married and joined the US Army when it was time to go off to war. He went to England until that war was over, and he got as far as Manila en route to the Japanese war before it ended. He became an Army civilian and based for a while in Utah, his wife Vivian's home state. His desire to go abroad led him to apply for and get a job with the CIA. The CIA sent him to Frankfurt, Germany, where he was content, except for the tragic and sudden death of Vivian during their first year overseas.

Velma met Werner and Vivian during their early days with the CIA in Germany, where Velma had gone to work for the CIA after an English divorce. He and she hardly spoke until sometime after his loss, however, once thrown together they eventually became close, probably because he learned about Velma taking golf lessons. They spent a year in Berlin before returning to Washington. Finally, they both received new assignments in Germany, where Werner first received an introduction to what became his association with the U-2 and A-12 programs.

After a few months in Wiesbaden, Weiss received an assignment to the U-2 base in Japan. He and Velma arranged to get married before he left so that she could join him when family quarters became available. They spent two years in Japan.

At Area 51, his contemporaries fondly referred to Weiss as the "Desert Fox." He started with OXCART from the beginning and served as the senior Central Intelligence Agency's officer throughout the program at Area 51 and its extension to Kadena. Weiss, a simple, however, a complicated man willing to take a centipede outside and release it, yet, he displayed on his office wall a bullwhip given him by his coworkers.

Weiss used Jim Freedman, one of the author's contemporaries with special projects to pick up his daily dispatches to CIA headquarters at Langley. Freedman delivered them to a worker at the McCarran Airport in Las Vegas. The following morning, Freedman dropped by the airport to pick up Langley's dispatches for delivery to Werner Weiss.

During this period, Werner continually demonstrated outstanding professional skill and initiative in managing the insurmountable tasks of maintaining base and project security. He was responsible for staffing, coordination, labor union relations, transportation, housing, and liaison with officials at local, state, and national levels.

His attention to detail in support of the aircraft operational and maintenance requirements contributed materially to the success of our mission. He oversaw all support activities for the A-12 first flight.

While labor union strikes occurred annually at the nearby Nevada Test Site, Area 51 experienced no labor strikes under Werner's watch.

Werner prided himself in the dining facilities at Groom Lake. The high quality and wide assortment of food served at Area 51 became something that everyone remembered about working at Area 51. The head of the mess hall, a black man, named Murphy Green is probably the one individual most known by those working at Area 51. He remained there while his customers came and went, everyone, remembering him for his iron fist way of running the mess hall. It did not matter who you were, what your station in life, you followed Murphy Green's rules in his mess hall. A great man respected and remembered by cadre and customers alike. Many still remember Murphy Green for his saying, "Preciate it, preciate it" each time someone paid at the cash register.

One must bear in mind that operations at the Area occurred around the clock and that support facility operated during the entire period with minor exceptions. This included hobby shops and many different support activities. The theater, rod, and gun club, swimming pool, bowling alley and much more all provided the necessary environment for this remote location. All operated under Weiss' control.

The North American Air Defense Command at Area 51

The North American Air Defense Command established procedures to prevent their radar stations from reporting the appearance of high-performance aircraft on their radar scopes. January 1962

Most of the new positions reflected the needs to maintain the support aircraft. A January 1962, agreement with the Federal Aviation Agency expanded the restricted airspace near Area 51 to prevent unauthorized aircraft overflights. The CIA cleared individual FAA air traffic controllers for the OXCART project to ensure that the unauthorized aircraft did not violate the order.

The CIA expanded its restricted airspace in subsequent years as the A-12 began flying longer training routes over the continental United States. The expanded airspace restriction allowed the A-12 to climb out to operating altitudes unobserved by other aircraft. The A-12 flew 2,850 such flights out of Area 51 during the OXCART phase before going operational with Operation BLACK SHIELD.

The North American Air Defense Command received briefings to prepare them for the flights of the A-12 and to preclude air defense fighters reacting to the appearance of high speed, high-flying unidentified aircraft. The CIA established procedures with NORAD to prevent A-12 flights initiating air defense alerts and to avoid air defense system radar stations reporting or discussing the appearance on their radar scopes of this high-performance aircraft.

NASA and the Area 51 Cover Story

The 1129th had a command post like other air force units, whereby a units' in-flight aircraft maintained contact with the base while airborne. In addition to the unit's aircraft, the command post also monitored radio transmissions of the military, commercial or private aircraft which were in the vicinity while the A 12 was airborne.

Most of takeoffs for the A-12 was to the north. During the climb out, the flight path crossed the heavily traveled commercial airways between Salt Lake City and San Francisco. On occasion, either one of the pilots of the commercial jets would see the A-12 climbing out, and all kind of chatter would go out on the airwaves. Such utterances as "Holy Cow did you see that!" Sometimes with verbiage not for innocent ears.

When this occurred, the word went out from the command post to a CIA office. Someone would meet the plane and swear the crew to secrecy under threat of being sent to ATTU in the Alaskan Island chain if they ever again uttered a word of what they thought they saw.

These sightings also occurred by Navy pilots from Fallon Navy Air Station flying to and from a gunnery range very close to the area, and they also would go into the broadcast mode stating what they had just seen.

The command post would advise them to go silent and return to base, and then contact the Oxcart cleared Base Commander to follow up what the CIA had advised the pilots to do. Area 51 would then launch a Cessna to the nearby Fallon NAS, which was close to the Area, meet the pilots, have them sign a paper stating that they never saw what they had just seen, and threaten them with severe penalties if they ever uttered another word about the incident.

On a humorous note, the Navy pilots knew someone other than them was flying something fast. Even though the mission planners at Area 51 made every effort to avoid the Supersonic A-12 flying over the populated area, they could not help but hit a few residential establishments with the sonic boom of the secret A-12. If anyone complained, the Navy at Fallon had to take the blame.

On 12 June 1962, NASA Deputy Administrator Dr. Hugh S. Dryden returned from an extended stay in Geneva, where he participated in international discussions with the Russians on the peaceful uses of outer space. He met with Messrs. Eugene Kiefer from the National Reconnaissance Office, Dick Bissell from CIA, and James A. Cunningham, Jr., director of administration DPS/DCI, and Acting Chief at DFD where he received an update on basic changes in the OXCART cover story.

Mr. Kiefer gave Dr. Dryden a brief, oral recapitulation of the progress to date in the flight test area of Project OXCART. Dr. Dryden suggested they reconsider the wisdom of identifying the OXCART vehicle as a "satellite launching system" for the simple reason that this statement carried the responsibility of explaining what kind of satellite this aircraft was capable of launching. In his view, mentioning satellites in the OXCART context opened a Pandora's box. He suggested the cover story reference the OXCART vehicle as an ALBM capability since all nations seemed interested in increasing the range of air-launched ballistic missiles.

Bissell disagreed with this cover story as it associated the OXCART aircraft with offensive weaponry to the degree that could well be inconsistent with a subsequent disavowal of any hostile purpose of this system. All parties felt concern about the cover story playing in the press, fearing the media reporting the US designing a covert launch capability for military hardware and clandestine satellite reconnaissance.

Refueling concepts required prepositioning of vast quantities of fuel at certain points outside the United States. The OXCART participants programmed special tank farms in California, Eielson AFB, Alaska, Thule afterburner, Greenland, Kadena Air Base, Okinawa, and Adana, Turkey. Since the A-12 used a specially refined fuel, they reserved these tank farms exclusively for use by the OXCART Program. A small detachment of technicians at these locations maintained the fuel storage facility and arranged for periodic quality control fuel tests.

At the Lockheed Burbank plant, Aircraft No. 1, known as Article 121, received its final tests and checkout during January and February 1962 and was partially disassembled for shipment to the site.

It became apparent very early in the OXCART planning that because of security problems and the inadequately short runway, the A-12 could not fly from Burbank. Lockheed and CIA successfully moved the full-scale radar test model in November 1959, as described above.

A thorough survey of the route in June 1961, ascertained the hazards and problems of moving the actual aircraft. The results showed they could transport a package measuring 35 feet wide and 105 feet long without significant difficulty if they removed obstructing road signs, trimmed trees, and leveled some roadsides.

The CIA and Lockheed planners made appropriate arrangements with police authorities and local officials to accomplish the safe transport of the aircraft. Lockheed furnished the special trucks, and the CIA provided security. Lockheed crated, loaded, and covered an entire fuselage, minus wings on the special-design trailer, which cost about $100,000. On 26 February 1962, it departed Burbank and arrived at Area 51 per plan.

The A-12 refueling requirements called for the CIA prepositioning special tank farms with vast quantities of its, especially refined and exclusive fuel outside the United States. The CIA and the air force established special tank farms at Beale AFB, California, Eielson, AB Kadena Air Base, Okinawa, Alaska, Thule Air Base, Greenland, and Adana Turkey. Very small detachments of technicians installed at these locations maintained the fuel storage facility and arranged for periodic quality control fuel tests.

Boomer

Probably the worst job of the Area 51 OXCART era was the assignment of a Lockheed individual known to the pilots as "Boomer." For several weeks, Boomer lived in a small camper, all alone, on the Current dry lake near Wells, Nevada. Wells is approximately 50 miles east of Elko, Nevada. Boomer's job was to measure sonic booms to see if the sonic boom of the A-12 differed from a conventual military jet. His only company was the pilots flying overhead in an F-101 during weather flights out of Area 51.

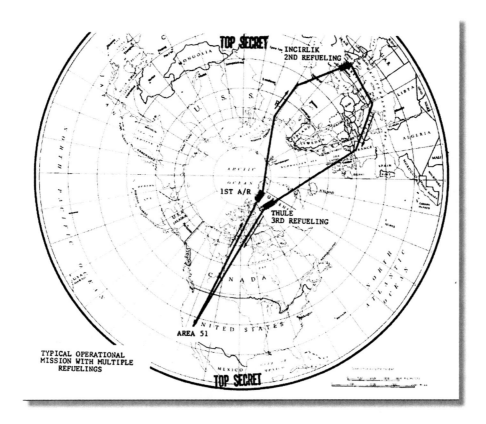

TYPICAL OPERATIONAL MISSION WITH MULTIPLE REFUELINGS

First Flight of the A-12

During January and February 1962, the Lockheed Burbank plant completed assembly off Aircraft No. 1 (serially numbered 121) and finished its final tests and checkout. Upon completion of its engineering tests, the company partially disassembled the aircraft for shipment to Area 51.

The CIA and the air force knew early in the OXCART planning that the existing runway at Area 51 was inadequate for landing the A-12 if flown from the Burbank manufacturing facility. The CIA and Lockheed successfully moved the full-scale anti-radar test model via a specially designed trailer truck over 453-miles of highway in November 1959.

A thorough survey of the route was made in June 1961, to determine the hazards and problems of moving the actual aircraft. It was found possible to haul a package measuring 35 feet wide and 105 feet long, without significant difficulty.

Before the move, Lockheed removed obstructing road signs, trimmed trees, and leveled roadside earth banks as required. The move required making appropriate arrangements with police authorities and local officials to accomplish the safe transport of the aircraft.

Lockheed boxed and covered the entire fuselage of the first A-12, known as Article 121, minus wings and loaded the crates on the carriage for hauling to Area 51 where the aircraft would begin ground runs and taxi checkouts before flight test.

Article 121, built and ground tested in Burbank during January and February 1962 was too secret to fly to the test site. Unlike the U-2 planes delivered from Burbank to Area 51 in a C-124, the A-12 was too large to transport on a cargo plane. During the night of 26 February, a specially designed trailer truck departed Burbank loaded with a huge crate measuring 35 feet wide and 105 feet long. The Lockheed-owned truck hauling the disassembled aircraft's fuselage departed the Skunk Works for the two-day trip to Area 51 with California and Nevada highway patrol escorts and CIA security officers. Hauled disassembled and in boxes to its Nevada test site, the A-12 posed a significant traffic hazard.

Leaking fuel became an unexpected problem the first time they filled the tanks in the A-12. Fuel leaked dangerously as the tank sealing compounds failed due to non-adherence to the metal. Lockheed engineers attempted to fix the problem by stripping the tanks of the faulty sealing compounds and relined them with new materials, once again delaying the first flight. Lockheed gave up on stopping the leaks and resorted to a fuel tank sealing compound that it deemed satisfactory for early stage flight testing and repairing the aircraft's fuel tanks.

On 25 April 1962, Lockheed test pilot Louis Schalk flew the A-12's first flight—unofficial and unannounced in keeping with a Lockheed tradition. The flight almost caused the loss of the only OXCART aircraft built so far. Schalk flew the plane less than two miles, reaching an altitude of about 20 feet because of severe wobbling that Johnson described as lateral oscillations caused by some navigation controls being improperly hooked up. Schalk, instead of circling and landing, put the plane down on the lakebed beyond the end of the runway in a cloud of dust and dirt.

The next day, Schalk flew the second flight that lasted about 40 minutes with the landing gear down, just in case. After a perfect takeoff, Schalk took the A-12 to about 300 feet, where the plane experienced shredding of the fillets of titanium. (Lockheed solved this problem on later aircraft by pairing triangular inserts made of a radar-absorbing composite material with the fillets.)

This also happened to Jack Layton during a training flight. He was refueling with a tanker when the boom operator said, *"Sir, I hope you are not planning to fly fast. I can see the ground through your wing."*

After four days of finding and reattaching the pieces, Article 121, just under one year later than originally planned, rolled out for its first official flight on 30 April 1962. Witnessing the event was an entourage of VIPs arriving on the Lockheed CL-329 JetStar corporate jet, a prototype and one of four of the planes ever built. These were senior air force officers and CIA executives from headquarters that included Dr. Herbert Scoville, Deputy Director for Research, Richard Bissell from the earlier U-2 Project AQUATONE. Other VIP observers included Lockheed chairman Courtland Gross and NRO, National Reconnaissance Office, Joseph Charyk along with representatives from affiliated government agencies and participating contractors.

The A-12 aircraft, piloted by Lockheed test pilot, Mr. Louis Schalk, took off at 170 knots. The plane, with a gross weight of 72,000 pounds, climbed to 30,000 feet altitude while sustaining a top speed of 340 knots. The plane landed after 59 minutes of flight, with the pilot reporting the aircraft responded well and was extremely stable. Mr. Johnson of Lockheed said it was the smoothest official first flight of any aircraft he had designed or tested.

On 4 May 1962, Schalk flew Aircraft 121 to Mach 1.1, reporting only minor problems for both flights.

Mr. John A. McCone, who succeeded Mr. Dulles as director, Central Intelligence in November 1961, headed the CIA's project headquarters for OXCART. Project headquarters was formerly the Development Projects Division in DDP and became the Office of Special Activities under the deputy director for Research. He sent a message of congratulations to Kelly Johnson wishing him continued success. He said that with the initial flight accomplished, he intended to direct t headquarters to make every effort to expedite all aspects of the program, to attain an operational posture as soon as possible.

By now, the Pratt & Whitney J58 (JT11D 20 model) engine development program had entered an intensive ground endurance testing phase. It had completed four 50-hour preliminary endurance engine tests as part of its preparatory to flight qualification and had added additional facilities. The company got its production engines on a firm schedule by accelerating its development and endurance testing. It expected delivery of the first production engine in August.

Delivery of the CIA Fleet of Mach 3 Spy Planes

The year 1962 produced and delivered five A-12 planes to Area 51. Aircraft No. 122 arrived at Area 51 on 26 June, where it spent three months in anti-radar testing by EG&G special projects before engine installation and final assembly.

Aircraft No. 123 arrived in August and flew in October. The following month, the two-seat trainer A-12 arrived at Area 51 for training OXCART project pilots yet to arrive. The CIA intended to power it with the J58 engines but settled for the smaller J75 engine because of the J58 engine delivery delays and a desire to begin pilot training concurrently with flight test. In January 1963, the trainer flew its initial flight and became the only A-12 to remain powered by the J75 engine.

The fifth and final aircraft delivered in 1962 was No. 125, which arrived at the Area during December 1962. On one of the hauls, an oncoming bus grazed a crate. CIA security officer Tom Stanks, following CIA protocol, paid the bus driver $5,000 in cash to continue his trip without reporting the minor incident.

The year 1962 first witnessed the Soviet Union moving into Cuba, which prompted the CIA and air force to maintain a regular reconnaissance vigil over the island that in October discovered offensive missiles. After that, U-2 reconnaissance increased until an Agency U-2 flown by a Strategic Air Command pilot on a SAC directed reconnaissance mission over Cuba was shot down by a surface to air missile on 27 October.

Suddenly, the CIA and the air force faced the prospect of possibly being denied continued manned, high-altitude reconnaissance of Cuba. This possibility gave the OXCART program greater significance than ever. The A-12 achieving, an operational status became one of the most top national priorities.

The close of 1962 found the program with two aircraft in flight test status, Two J 75 engines powered on of the A-12s and one J75 and one J58 the other. The A-12 had reached a speed of Mach 2.16 and altitude of 60,000 at this point. The aircraft had not achieved design speed and altitude to access the auxiliary components arid payloads thus far flown at low speeds.

The flight test program was experiencing slow delivery of the engines and poor engine performance at altitude, which accounted for the slow start in the flight test program. Instead of Pratt & Whitney delivered the 14 engines promised by late November 1962, the company provided only nine with only four considered usable in the flight test program.

At a time of greatest need to reach an operating capability, the program appeared to be at its lowest ebb. The Development Projects Division Project Officer took the initiative to send a message to the president of United Aircraft Corporation to place pressure on its top management. In it, he said,

"Mr. Horner from Mr. McCone: 1 have received word that J58 engine deliveries have been delayed again due to engine control production problems. At this moment, we have four engines with satisfactory controls and one questionable to operate two aircraft. By the end of the year, it appears we will have barely enough J58 engines to support the flight test program: adequately. Furthermore, we have been forced to use J75 engines in airplanes one, three and four. Furthermore, due to various engine difficulties, we have not yet reached design speed and altitude. Engine thrust and fuel consumption, deficiencies at present prevent sustained flight at design conditions which is so necessary to complete development of the complete system. This situation gives me the greatest of concern because of the critical importance of the program. It is necessary that the United Aircraft Corporation, Pratt & Whitney, particularly the Hamilton Standard Division, place the highest priority on the solution of all technical and production problems and assign to them the most senior and competent men within your combined organizations. The OXCART program has been designated of the very highest national priority and cannot emphasize enough: the necessity of its reaching operational status as quickly as possible. Would appreciate your informing, me personally of your actions taken to correct these difficulties."

By the end of January 1963, Pratt & Whitney had ten engines available at Area 51, two of them making the first flight with J58 engines on 15 January 1963. After that, flight testing accelerated with all A-12 aircraft fitted with the J58 engines upon delivery to Area 51. The contractor personnel went to a three-shift workday.

The A-12s and the trainer lined up at Area 51

Flight Testing the A-12

As expected with a revolutionary new plane concept, new problems cropped up with each succeeding step into the higher Mach regime. The air inlet and its control system became the single most significant issue plaguing the flight development of the A-12. Lockheed fixed the problem by designing a supersonic inlet or air induction system to provide the best possible aerodynamic performance over a range of supersonic Mach numbers. The redesign provided a stable and steady flow of air to the engine, eliminating a design problem that the Air Force's SR-71 later never experienced.

Advancing flight testing to the Mach 2.4 to 2.8 range caused the aircraft to experience severe inlet duct roughness. This was where an improper airflow match occurred between the inlet and the engine. Lockheed also determined a contributor to the roughness being improper aerodynamic contouring of the inlet duct, which redesign corrected.

The control system activating movement of the inlet spike caused mass flow rate changes in the engine duct. In a phenomenon known as an unstart, the intake's capture plane did not match the downstream mass flow, causing the engine to lose thrust, causing a violent breakdown of the supersonic airflow that resulted in a violent, temporary loss of control until the intake restarted. To resolve the problem, Lockheed developed two control designs, one hydromechanical and the other, an electronic backup approach. Solving the unstart problem proved so difficult it prolonged the date when the CIA could declare the A-12 operationally ready.

During the spring of 1963, Lockheed encountered the concerning and costly problem of workers leaving nuts, bolts, clamps, and other debris in the manufacturing process. Foreign objects left behind were sucked out of the nacelle's inner nooks and crannies during ground run-up or takeoff and entered the intake duct to be ingested by the engines to cause significant damage to the engine.

A thorough investigation eliminated the possibility of sabotage and attributed the problem to carelessness and poor housekeeping that rigorous inspections and controls corrected.

Chapter 7 – Article 123 Down

The CIA lost its first aircraft while on a routine training flight in May 1963 when Ken Collins, ejected from the aircraft after obtaining an erroneous airspeed indication and. Collins was unhurt when the aircraft crashed 14-miles south of Wendover, Utah.

For public consumption, the CIA identified the A-12 Article 123 as an F-105. The CIA and air force crews from Area 51 removed the wreckage from the scene and returned it to Area 51 by 26 May. Agency security at Area 51 identified all individuals at the crash scene and requested they sign secrecy agreements. Press, inquiries concerning the pilot's identity were forthcoming, and air force sources issued the story that the F-105 aircraft was on bailment to Hughes Aircraft Company from Air Force Logistics Command, Wright-Patterson Air Force Base, Ohio. The cover story went on to say that the aircraft piloted by Mr. Collins originated from Wright Patterson Air Force Base on Friday. Following a brief stop at Nellis AFB, in the early Friday afternoon Collins, while testing a classified system, experienced aircraft difficulties and bailed out of his aircraft near Wendover, Utah.

The CIA grounded all A-12 aircraft for a week pending an accident investigation that found ice plugging the pitot-static tube, an open-ended right-angled tube pointing into the flow of a fluid and used to measure pressure. Ice filling the pitot tube that measured the airspeed of the aircraft had caused the faulty cockpit instrument indications.

Air Force A-12 procurement tightly interwoven with the chronicle of OXCART development and events occurring on its periphery. In December 1960, the air force, using project procurement channels, contracted for three long-range interceptors (LRI) versions of the A-12 that used the cryptonym KEDLOCK. In January 1962, the air force added another five A-12 buy to the existing A-12 production contract termed the KEDLOCK program.

In 1962, the deputy director of Research and Engineering, Department of Defense proposed another reconnaissance concept, which offered an alternative to the manned supersonic reconnaissance system. It was a Mach 3.3 ramjet drone aircraft that launched from an A-12 mother aircraft identified at the M-21 and the drone, the D-21.

After completing feasibility studies by Lockheed and evaluation by the Pentagon and the National Reconnaissance Office on 17 October 1962, the NRO director authorized management and technical monitoring of development assigned to CIA (project headquarters, how OSA).

On 4 June 1963, for political reasons (during an air force-CIA tug of war on the NRO), the responsibility was shifted from the CIA to the air force. The program became known as TAGBOARD.

The CIA scheduled two of the five KEDLOCK A-12 aircraft for conversion to a TAGBOARD drone launch configuration, and on 6 November 1963, transferred the remaining three to the OXCART program.

On 29 January 1963, the secretary of defense approved the purchase of six additional aircraft for air force use as a general-purpose reconnaissance vehicle configured to carry a variety of intelligence collection systems.

In August 1963, the secretary of defense approved an additional procurement of 25 A-12 aircraft in a new 71 configuration designated the R 12 (later redesignated as the SR 71 Blackbird). The CIA again acted as the procurement agent.

Crash site of A-12 Article 122 flown by Ken Collins

The YF-12 Project Kedlock

In October 1962, the Air Force needed to replace the canceled F-108A Rapier and saw a potential of developing a Mach 3 interceptor with a variant of the CIA's A-12. The Air Force ordered three modified A-12s, first designated the AF-12 and then the YF-12A. Lockheed designed and built three modified A-12s under a project codenamed KEDLOCK.

Lockheed removed the camera from the Q-bay and installed a missile launch cockpit with the necessary life support for a second crewman for the fire control system. Hughes added a radar system in the nose and pylons for three air-to-air missiles. The aircraft's mission was to intercept the new Soviet supersonic bombers long before they reached the United States.

The Air Force planned for a fleet of as many as 100 but built only three that Lockheed delivered to Area 51 during 1963-64.

The CIA was involved with the project only in giving up three A-12 airframes and helping write "black" contracts.

The President's Announcement

A week after taking office in 1963, President Johnson received a briefing on Area 51 and the development of the Mach 3 Blackbird family. He directed the preparation of an announcement for the spring of 1964. Then at his press conference on 24 February, he read a statement of which the first paragraph was as follows:

"The United States has successfully developed and tested an advanced experimental jet aircraft, the A-11, in sustained flight at more than 2,000 miles per hour and altitudes more than 70,000 feet. The performance of the A-11 far exceeds that of any other aircraft in the world today. The development of this aircraft has been made possible by major advances in aircraft technology of great significance for both military and commercial applications. Several A-11 aircraft are now being flight tested at Edwards Air Force Base in California. The existence of this program is being disclosed today to permit the orderly exploitation of this advanced technology in our military and commercial program."

The president went on to mention the "mastery of the metallurgy and fabrication of titanium metal. He credited Lockheed and Pratt & Whitney and remarked of keeping appropriate members of the senate and house informed. He prescribed keeping the detailed performance of the A-11 classified.

The President's reference to the "A-11" was of course deliberate. "A-11" had been the original design designation for the all-metal aircraft first proposed by Lockheed; subsequently it became the design designation for the Air Force YF-12A interceptor which differed from its parent mainly in that it carried a second man for launching air-to-air missiles. To preserve the distinction between the A-11 and the A-12, Security had briefed practically all participating personnel in government and industry on the impending announcement. OXCART secrecy continued in effect. There was considerable speculation about an agency role in the A-11 development. However, the government never acknowledged it. News headlines ranged from "US has dozens of A-11 jets already flying" to "Secret of sizzling new plane probably history's best kept."

Initially, the president wanted to announce the CIA's A-12. The CIA refused and told the president that if he wanted to out the Blackbirds, he could do the Air Force's YF-12. The president remained persistent about showing off the Blackbirds and agreed to announce the YF-12 and not mention the CIA's secret Area 51.

Instead, the president said that "the A-11 aircraft was at Edwards Air Force Base undergoing extensive tests to determine their capabilities as long-range interceptors." It was true that the air force in October 1960, had contracted for three interceptor versions of the A-12, and they were by this time available. However, now when the president spoke, there were no A-11s at Edwards, and there had never been.

Project officials knew that he was about to make the public announcement, but did not know exactly when. Caught by surprise, they hastily flew two air force YF-12As to Edwards to support the president's statement. So rushed was this operation, so speedily were the aircraft put into hangars upon arrival, that heat from them activated the hangar sprinkler system, dousing the reception team which awaited them.

Thenceforth, while the OXCART continued its secret career at its site, the A-11 performed at Edwards Air Force Base in a high glare of publicity. Pictures of the aircraft appeared in the press; correspondents could look at it and marvel, stories written. Virtually no details were made available, but the technical journals nevertheless had a field day. The unclassified Air Force and Space Digest, for example, published a lengthy article in its issue of April 1964, commencing: "The official pictures and statements tell very little about the A-11. However, the technical literature from open sources, when carefully interpreted, tells a good deal about what it could and, more importantly, what it could not be.

The YF-12A was quite similar in overall configuration to the A-12 from which it derived. It differed from the A-12 primarily in having a second crewman in a position immediately behind the pilot. This added a second crewman to operate the extremely powerful and capable Hughes AN/ASG-18 pulse Doppler fire control radar originally developed for the F-108 Rapier. The nose of the YF-12 contained the AN/ASG-18, with the forward chines cut back to accommodate the 40-inch radome. The ASG-18 radar supposedly had a search range as high as 500 miles. The forward edges of the cut-back chines contained infrared sensors.

The YF-12A also differed from the A-12 in having armament. This armament consisted of four Hughes AIM-47A Falcon air-to-air missiles housed internally in chine bays previously used to carry the reconnaissance equipment. The AIM-47A was originally known as the GAR-9 and (like the ASG-18 radar) was originally intended for the F-108 Rapier. When fired, the Falcon missiles explosively ejected from their bays, and their rocket motors fired. Powered by a storable-propellant liquid-fueled rocket, the AIM-47A had a maximum speed of Mach 6 and an interception range of 115 miles. It had a launch weight of about 800 pounds. The missile relied on semi active radar homing for midcourse guidance to the immediate vicinity of the target, homing in on reflections off the target resulting from transmissions from the huge ASG-18 radar. However, it used terminal infrared homing for the final run into the target. The AIM-47 could carry a 250-kiloton nuclear warhead.

Secretary of Defense Robert McNamara canceled KEDLOCK in early 1968 as a cost-cutting measure. Thus the aircraft never deployed operationally. The Air Force bore all the costs of the YF-12A superseded by the F-111. Two of the planes went to the National Aeronautics and Space Administration for research, and one became a trainer for the SR-71 program.

Although it yielded large amounts of research data, NASA terminated the YF-12 program in the late 1970's when its research agenda shifted from speed to efficiency. During its nine-year life, the YF-12 research agenda logged 297 flights and approximately 450 flight hours (including hours in a SR-71A designated YF-12C.

The YF-12

Project Tagboard

In October 1962, the CIA, in a project codenamed TAGBOARD, authorized the Skunk Works to study the feasibility of modifying the A-12 to carry and deploy a reconnaissance drone for unmanned overflight of denied areas.

The two M-21s, modified A-12s, carried and launched the expendable 10-ton D-21 unmanned reconnaissance drone. The main difference between the A-12 and M-21 models was that the M-21 had the sensor package in the Q-bay behind the pilot's cockpit removed to make way for a second seat for the Launch Control Officer (LCO).

They designated the mother ship, the M-21 to avoid confusion with the A-12. The D-21 drone was 43 feet long, weighed over five tons, had a ramjet engine, could reach a speed of over Mach 3.3 at 90,000 feet, fly over 3,000 miles, and had the smallest RCS of anything Lockheed had yet designed. The D-21 looked like a half-scale single engine Blackbird, albeit without a cockpit, and was powered by a Marquardt ramjet. Because the ramjet power plant only worked at speeds above Mach 3, it was necessary to first get the D-21 up to that speed on the back of a Blackbird before the ramjet could power itself.

The drones launched well away from targets, flew their missions, and returned to a preprogrammed location in international waters. There they jettisoned a payload called a package that a C-130 snagged in midair. The drone then self-destructed with a barometrically activated explosive device.

In June 1963, the air force had overall charge of unmanned reconnaissance aircraft, making the drone its project.

Lockheed eventually built two M-21s and 38 drones. Lockheed test pilot Bill Park flew all the M-21 flights. On 30 July 1966, the fourth TAGBOARD test, a launch mishap caused the drone to strike the mother ship, causing it to crash, killing LCO Ray Torrick and prompting Kelly Johnson to end the program. The Air Force continued the program using B-52s to launch the drones against Communist Chinese targets in a project called SENIOR BOWL. The Air Force flew four missions starting in November 1969 and canceled the program in July 1971 for failure to complete a successful test or mission.

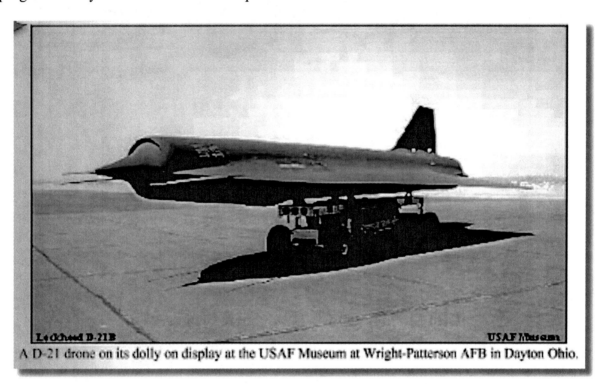
A D-21 drone on its dolly on display at the USAF Museum at Wright-Patterson AFB in Dayton Ohio.

The A-12 modified as the mother ship (M-21) for the stealthy D-21 drone to spy on denied territories.

LBJ Surfaces the A-11 Aircraft Program

By January 1953, there was an increasing number of the A-12 family of aircraft on order or proposed for order. The air force getting into the Mach 3 interceptor business with the YF-12 and the Mach 3 reconnaissance business with the SR-71 made it difficult for the CIA to preserve the covert character of the OXCART program. Compounding the problem was Secretary of Defense Robert McNamara giving serious consideration to issuing a press release announcing the existence of an X 21 development (the AF version of the A-12) in January 1963.

Since the action tied directly to the XB 70 controversy, the CIA feared more political debate and critical examination possibly exposing the A-12.

When the president's Foreign Intelligence Advisory Board reviewed the situation, it was the Board's conviction that it would be best not to make any statements which would possibly result in exposing A-12 development, or any military version thereof.

The president decided not to surface the R 12. While the White House did not make a statement at that time, but it was apparent to the CIA that public disclosure public disclosure of Lockheed's work in Mach 3.0 aircraft manufacturing was inevitable as the delivery date of the first R 12 approached.

The agency's chief concern about the president surfacing the Blackbirds was how to protect, the anti-radar aspects and capabilities of the A-12, and protecting the scope and nature of the CIA's activities at Area 51

Thus far, the CIA's development, construction, and a year of flight testing and training in the OXCART program had avoided attracting public attention. However, with all the A-12 variants appearing on the scene, it now seemed advantageous to surface some version of the A-12 type aircraft to serve as cover for the A-12 in the event of an incident or inadvertent exposure.

The May 1963 loss of Aircraft No. 123 became a case in point as the matter continued under review during 1963. The Department of Defense had trouble in concealing its participation in the program due to the increasingly high rate of expenditures as the Strategic Air Command's R 12 aircraft came into production.

Political pressures increased in the DOD/Congressional controversy involving appropriations for the improved manned interceptor and the manned bomber. Also, the Administration desired to make Mach 3.0 aircraft technology data available to participants and evaluators of the XB-70 and SST programs.

There was a growing awareness in the industry and the press of the existence of the program. Operational activity, both in numbers of aircraft and extended flight profiles, was increasing the probability of random sightings and incidents.

All parties agreed that an announcement to the public was necessary. President Johnson was apprised of the matter, on 29 November 1963, a week after taking office. He decided that surfacing was in order and directed the preparation of an announcement to occur during the spring of 1964.

The president, at his 29 February 1964 press conference, made the following statement:

The United States has successfully developed an advanced experimental jet aircraft, the A-11, tested in sustained flight at more than 2,000-miles per hour and altitudes more than 70,000 feet. The performance of the A-11 far exceeds that of any other aircraft in the world today. The development of this aircraft has been made possible by major advances in aircraft technology of great significance for both military and commercial applications. Several A-11 aircraft are now being flight tested at Edwards Air Force Base in California. The existence of this program is being disclosed today to permit the orderly exploitation of this advanced technology in our military and commercial program.

This advanced experimental aircraft, capable of high speed and high-altitude and long-range performance of thousands-of-miles constituted a technical accomplishment that facilitated the achievement of some significant military and commercial requirements. The A-11 aircraft now at Edwards Air Force Base was undergoing extensive tests to determine their capabilities as long-range interceptors. The lessons learned from this A-11 program would greatly assist the development of supersonic commercial transport aircraft. For example, the mastery of the metallurgy and fabrication of titanium metal required for the high temperatures experienced by aircraft traveling at more than three times the speed of sound is one of the significant technical achievements of this project. Those directly engaged in the Supersonic Transport Program are arranging to make this and other important technical developments available under appropriate safeguards.

"This project started in 1959 had kept appropriate members of the Senate and the House fully informed of the program since the day of its inception, The Lockheed Aircraft Corporation of Burbank, California is the manufacturer of the aircraft.

The Pratt & Whitney Aircraft Division of the United Aircraft Corporation designed and built the aircraft engine, the J58. The Hughes Aircraft Company built the experimental fire control and air to air missile system for the A-11.

"Given the continuing importance of these developments to our national security, the detailed performance of the A-11 will remain strictly classified, and all individuals associated with the program directed to refrain from making, any further disclosure concerning this program. I do not expect to discuss all of you after this meeting. If you care, Mr. Salinger will carry out the appropriate arrangements."

The reference in the president's announcement to the then A-11 was deliberate. To announce the aircraft as being one of the "X" series would not have been completely true, whereas the "A-11" was the original designation of the all-metal aircraft first proposed by Lockheed. The air force subsequently redesignated the surfaced "A-11" as the YF 12A.

Project security preserved the distinction between the "A-11" and the A-12, by briefing practically all affiliated personnel in government and industry on the impending announcement. This was to preclude relaxation of the OXCART security standards, though there was considerable press speculation on an Agency role in the A-11 development that was never acknowledged by the Government.

The air force flew two A-11s from Area 51 to Edwards Air Force Base, California, where all subsequent A-11/YF 12A activity occurred in time to the president's announcement.

On 24 July 1964, the president announced the SR 71 (R 12) development. Fortunately, the announcement preserved the existence of Area 51, its activities, and the role of the CIA in its operation.

Building up Area 51

In 1963, the CIA needed to accommodate the additional aircraft ordered (three YF 12's and five more A-12s). It did so by expanding its facilities at Area 51 in 1963 with the construction of an additional hangar, a new mess hall, an administration building, and a special handling building for the camera systems. Area 51 facility population reaching 1,423 in November 1963 required the addition of BOQ facilities.

The agency had gained enough experience in the operation of the aircraft that it decided to extend the runway. It needed to provide an acceptable margin of safety in the event of rejected take off that would allow the plane to use the additional length to come to a stop rather than run out on the lakebed where structural damage might occur due to surface roughness.

The agency extended the runway by 11,000 feet of load-bearing asphalt with additional 5,000 feet of graded overruns on each end.

Flight Testing Continues

By the end of 1963, the A-12 had flown 573 flights, totaling 765 hours since the first flight in April 1962. It had nine aircraft in the inventory, and three becoming available by the end of March 1964. On 20 July, flight test aircraft had achieved the first Mach 3.0 flight.

In November, the A-12 had reached Mach 3.2 at 78,000 feet altitude. The inlet duct roughness, the problem appeared to be solved, but the inlet performance was still below design requirements. The entire aircraft system operated reasonably well within the test limits, thus far, Lockheed was developing a jamming device designed to defeat the guidance link of the SA 2 missile system and expected to test it shortly. The two camera systems performed satisfactorily within the limits of airspeed and altitude flown to date.

The three KEDLOCK aircraft inherited a third camera system by Hycon that is now part of the OXCART sensor inventory.

Lockheed directed maximum effort for the next few months toward optimizing the inlet to a firm production configuration. The Skunk Works engineers strived to improve aircraft transonic performance with the aircraft operating at high temperatures and altitudes to provide the proper environment in which to test equipment and sensors.

On 3 February 1964, the agency conducted the longest sustained flight at design conditions on a flight lasting 10 minutes at Mach 3.2 and 83,000 feet. Flight test data continued showing the inlet performance as deficient between Mach 1.8 and 3.2, resulting in excessive fuel consumption.

In May 1964, the CIA formed a special task force of senior performance personnel from the contractors involved to focus exclusively on the aircraft inlet/propulsion interface problems. The task force stayed in residence at Burbank until it developed a comprehensive inlet/engine improvement program.

By May 1964, the OXCART program had reached a limited Mach 2. 35 operational capability, but the basic inlet/propulsion interface problems remained. Project SKYLARK

At this stage of Project OXCART, operational planners had in place a set program to develop flight planning based on performance data as it became documented and coordinated with the flight test and Detachment training. The CIA added each new performance plateau reached and proven to this program,

The rush was on to get the A-12 operational to replace the vulnerable U-2 project headquarters foresaw a possible contingency whereby the A-12 might need to overfly Cuba. The agency began in-house operations planning for that purpose designated as Operation SKYLARK,

Meanwhile, in mid-year of 1964, the flight testing and detachment training suffered a serious interruption when on 9 July 1964, Lockheed test pilot Bill Park crashed Article 133 just short of the runway on its final approach during a shakedown flight. The aircraft had descended to 500 feet altitude and was at an airspeed of 200 knots when it entered a smooth, steady roll to the left. Park could not overcome the roll and rode the plane into an approximate 45-degree bank angle. At 200 feet altitude, Park ejected safely. The ensuing crash investigation found the right outboard elevon servo valve stuck in a partially open position as the cause of the accident. Consequently, the CIA grounded the fleet for a month for Lockheed to incorporate fixes,

The Park crash reduced the fleet to eleven aircraft. Because the accident occurred inside the Area 51 reservation, the accident avoided any publicity. However, the CIA admonished all Area 51 personnel not to discuss the accident. An accident investigation-board quietly investigated the crash so that neither the press nor the public ever knew about it,

The Soviet buildup in Cuba continued and feared the shoot down of the U-2 over Cuba, SKYLARK, on 11 August, became a directed project with the imposition of an emergency operational readiness date of 5 November. Lockheed completed the delivery of the three A-12s in March totaling 13. However, the loss of Aircraft 123 and 133 reduced that total to eleven,

The CIA's need for the A-12 to become operational became an exerted and strenuous effort for the plane, pilots, and supporting elements. The agency reached the desired capability by setting a goal of Mach 2.8 and 80,000 feet altitude, a goal that resulted in 16 necessary changes or additions to the aircraft. Still, there were only one of the electronic countermeasures systems available by the CIA's readiness date. Furthermore, camera performance needed validation at the SKYLARK level, the pilots Mach 2.8 qualified, and the necessary coordination with supporting elements accomplished,

A senior intra-governmental committee that included a representative from the president's Scientific Advisory Committee (PSAC) had examined the problems inherent to overflying the Soviet activities occurring in Cuba. After assessing the situation, the Committee decided the CIA could fly the first few overflights without the full complement of defensive systems. Any overflights after that would require countermeasures,

At this point, the delivery schedule of ECM equipment was compatible with this course of action. The agency had to turn back four of the six Detachment aircraft for Lockheed technicians to modify and update. By 19 October, Lockheed had completed the necessary hardware changes and returned the plane to the detachment more than two weeks past the scheduled date because of various aircraft systems integration problems. A multitude of unrelated malfunctions required correction to make the aircraft acceptable from an operational reliability standpoint. Having to correct these problems, in turn, delayed everything, the detachment pilot training, systems, the payload validation, and the overall determination as to reliability. Worse yet, because of the delayed delivery of aircraft, the agency could qualify only three of the project pilots for the contingency,

The detachment prepared for operational flights over Cuba by simulating SKYLARK missions during training flights. It practiced multiple aerial refueling and operating systems and payloads, and on 5 November 1964, the CIA announced a limited emergency SKYLARK capability. With a two weeks' notice, the OXCART Detachment could accomplish a Cuban overflight. However, the CIA must do it with fewer ready aircraft and pilots than originally planned,

At Area 51, the agency devoted primary Detachment emphasis in the ensuing weeks on developing a sustained SKYLARK capability of five ready pilots and five operational aircraft. Mainly, they determined aircraft range and fuel consumption. The pilots attained repeatable, reliable operation as they completed pilot training. Mission planners prepared a family of SKYLARK missions with complete coordination of routes with NORAD, CONAD, and FAA. The CIA and the 1129th SAS air force personnel exercised command and control through operational readiness inspections and command post exercises,

Flight Test Achievements

In the weeks that followed, the Lockheed flight test program placed primary emphasis on accelerated testing of defensive systems. It sought to attain the original Mach 3.2 performance specifications that resulted in improved equipment and component reliability. They saw a high speed of Mach 3.27, achieved an altitude of 83,000 feet, and a sustained flight of 32 minutes over Mach 3.0 at 82,000 feet,

Three years' experience in high-altitude, high speed flight testing had proven that achieving a reliable capability at design specifications a challenging and frustrating task. It took six months of flying for the A-12 to reach Mach 2 and 15 months to reach Mach 3. Two years after first flight, the A-12's Mach 2 time totaled 38 hours. The Mach 2.6 speed totaled three hours and speeds at or above Mach 3 less than one hour. After three years, the A-12's Mach 2 time had increased to 60 hours, Mach 2, six times to 33 hours, and Mach 3.0 time to 9 hours with all Mach 3.0 time confined to Article 121, the flight test aircraft, and Detachment aircraft restricted to Mach 2.9,

Reaching Operational Capability - 1965

From the first flight on 26 April 1962 and through 31 December 1964, all thirteen aircraft built (eleven of which remained) had flown a total of 1160 flights, and 1616 hours. Twin J58 engines had powered 743 flights and had accumulated 928 hours. Four of the 11 A-12s were still in flight test and seven aircraft, including the trainer, now assigned to the detachment. The SKYLARK aircraft, namely Articles 125, 127, 128, 132, and later, 129 and 131 had undergone inlet modifications and a faster climb schedule that now allowed a Mach 3.05 speed and an increased range of 2,500 nautical miles. The A-12's operational performance was now Mach 2. 9, with a range of 1700 nautical miles, and an altitude of 76,000 feet,

Project SILVER JAVELIN – The OXCART Validation Process.

On. 27 January 1965, in a mission codenamed SILVER JAVELIN, the flight test aircraft flew the first in a series of long-range, high-speed flights. The flight duration was 1.40 hours, 1.15 hours of it above Mach 3.1. The total range was 2,580 nautical miles, which met the mission's purpose of demonstrating the maximum range capability. The plane cruised at altitudes between 75,600 and 80,800 feet. The mission represented the longest sustained flight closely approximating design conditions.

Exercise ECHO – a Simulated SKYLARK Mission.

On 28 January 1965, project headquarters, SAC, Detachment, and FAA jointly participated in Exercise ECHO to simulate a SKYLARK mission. The exercise served to involve all elements that, participate in an actual overflight. All phases of the joint event proved significantly successful in that it inaugurated the detachment's operational mission type training,

SKYLARK Modifications.

The OXCART project embarked on two modification programs at Area 51 in 1965 with the objectives of increasing speed, improving reliability and range, increasing mission duration, and incorporating electronic countermeasures. The first modification began in the spring of 1965 to improve the SKYLARK capability to a Mach 3.05 level. Other modifications included further improvements in the inlet system, strengthening the structure of plastic panels, increasing the aircraft's supply of inert fuel (nitrogen), and the tank's pressurizing medium to provide for more refueling on a mission, strengthening the rudder pose and incorporating minor equipment changes,

The MMM Program.

The MMM program included changes that allowed a faster climb schedule. It provided space in the chines to install the ECM equipment, which required strengthening fuselage station joint 715 because of the increased bending movement introduced by the weight of the ECM gear. The modifications included increasing the liquid oxygen supply to allow for longer missions,

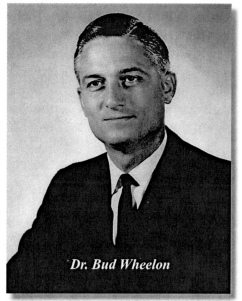

Dr. Bud Wheelon

The MMM program revised the inlet control system to improve the performance of the primary inlet control system where inlet roughness and shock expulsion had permitted speed extensions only in small increments to Mach 2 8. At the insistence of CIA's Deputy Director of Science and Technology, Dr. Albert D. "Bud" Wheelon, Kelly Johnson opted for the backup electronic inlet control system over the hydromechanical system,

To his credit, Kelly Johnson phased the modification programs in such a way that always made at least five aircraft available to the detachment for training and proficiency flying. Once the detachment accepted the modified aircraft, the pilots began acquiring experience in the higher Mach numbers that the modifications permitted.

Mainly, the MMM program standardized the configuration of the A-12 aircraft where the numerous approaches to solving A-12 problems had resulted in nonstandard configurations for several of the planes. At completion, the modification programs provided a uniform operating fleet of A-12s, equipped and capable of performing the design mission. The changes proved fruitful when on 25 March 1965, a project pilot first flew the A-12 aircraft at Mach 3.0. After that, the pilots routinely accumulated time at the high speeds to Mach 3 qualified all the detachment pilots by mid-1965,

The High Point of Project OXCART.

The year 1965 became the culmination in the level of activity in the history of the OXCART program with the CIA's flight test facility at Area 51 reaching full physical growth. With the construction now completed, eight Butler shelter type hangars now housed the detachment aircraft. Area 51 finally had commercial power and was now comparable, to a reduced level of a typical US Air Force installation,

The prime contractors working three shifts a day on the modification programs at Area 51 had increased the population to 1,835 with the air force's 1129th SAS Detachment T/O now at 280 personnel because of the TAGBOARD activity began in August 1964,

Three Constellation aircraft how made daily scheduled workday flights between Burbank and Area 51 to transport contractor, personnel, and vital freight. Two daily C-47 shuttle flights were now flying from Las Vegas to transport Edgerton, Germeshausen and Grier, Inc. (EG&G) personnel to operate the radar range. Formed in November of 1947 by Dr. Harold "Doc" Edgerton, Kenneth J. Germeshausen, and Herbert E. Grier three Massachusetts Institute of Technology scientists, EG&G had long been the prime contractor for atomic testing at the adjoining Nevada Proving Grounds ran by the Atomic Energy Commission (AEC). EG&G had moved its radar cross-section (RCS) facility from Indian Springs north of Las Vegas to Area 51 where its special projects team of engineers and technicians had and continued to perform the RCS evaluations of the A-12 for the CIA. The El Paso, Texas firm of Reynolds Electrical and Engineering Company (REECo), another major contractor at the Nevada Proving Grounds complemented the EG&G special projects team at Area 51 where it performed most of the infrastructure construction work.

One of the CIA's key people working with EG&G special projects was Gene Poteat, LL.D. an Electronics/Physicist/Missile Guidance Engineer. Poteat now the program manager for payloads and ECM for the U-2 AQUATONE and A-12 OXCART Programs in the Development Projects Division (DPD). (DPD became the Deputy Direct of Science and Technology, DDS&T under Dr. "Bud" Wheelon, Ph.D.) Early in the OXCART program, Poteat had run project Palladium to exploit the Soviet Union's new and highly feared TALLINN (NATO name: TALL KING) radar deployed to Cuba. Each TALLIN complex consisted of three launch sites, each with six launch positions and one radar. Ever since Project Palladium, the CIA knew because of this new radar, the A-12 could not overfly the Soviet Union as the U-2 had done,

Nonetheless, the CIA needed to develop an electronics countermeasure system to protect the A-12 while flying over Soviet-sponsored denied territory. Part of the effort was procuring a surplus the air force radar from Fort Fisher, North Carolina. The agency moved the system to Area 51 where the special projects team operated it to simulate the Soviet Union's FAN SONG radar during the ECM system development during the A-12's flight testing program,

Can the Soviet Union Track and Shoot Down the A-12?

The late 1950s became the heyday of the U-2 reconnaissance aircraft when it flew with impunity over the Soviet Union. The U-2 brought back the most-sought-after intelligence at the time, confirmation that no real bomber or ballistic missile gap existed between the US and the Soviet Union.

However, the U-2 also brought back something else: a foreshadowing of its impending demise. The U-2 camera, along with its rudimentary electronic intelligence (ELINT) receivers, picked up indications of a Soviet antiaircraft defense buildup. It revealed a new and better surface-to-air (SAM) missiles and radars used by the Soviets in attempts to shoot down a U-2. Neither Russian interceptor fighters nor its SAMs succeeded in a shoot down until 1 May 1960.

At the time of the shootdown, the CIA's U-2 program office was already well along in developing the U-2's replacement, the A-12 Archangel code-named OXCART at Lockheed's Skunk Works in Burbank, California. The A-12 Project OXCART flew at about 90,000 feet, at Mach 3.3 to become the predecessor to the air force's better-known SR-71 Blackbird.

Meanwhile, the CIA and the air force continued looking to the future, the need for reconnaissance over Russia. To this end, the CIA and the air force jointly initiated the ultimate observation system underway in a parallel development. This system became the CORONA satellite, the first in a long series that eventually replaces all overflights, including the A-12 Project OXCART.

Concerns about the vulnerability of the yet-to-fly OXCART to the evolving Soviet air defense network generated the basis for the most secret and sensitive aspect of the project. The A-12 Project OXCART called for making it invisible to the Soviet radars—the first-ever stealth aircraft

Parangosky served at Area 51 from 1956 through 1965 during the flight testing and certification of the CIA A-12 Project OXCART successor to the U-2. He served with the agency's Office of Special Activities as the program's executive officer and the program manager. His contributions paved the way for creating the CIA's Directorate of Science and Technology directed by Dr. Albert Dewell "Bud" Wheelon, Ph.D.

In 1962, Dr. Wheelon arrived at Area 51 and was immediately concerned about the vulnerability of the A-12 to Soviet air defenses. He assembled a small, however, very competent analytical unit to anticipate the threat to missions over the USSR. Using IBM computer, the CIA simulated SA-2 intercepts of the A-12 with precise accuracy.

The simulations showed OXCART at risk with altitude alone not enough to defeat the combination of powerful acquisition radars and radar-guided missiles.

The SA-2 missiles had enough thrust to reach OXCART operating altitudes. The CIA decided to ensure the A-12 survival by exploiting the missiles' lack of control at high-altitude by adding radar jammers to the aircraft

The second part of the CIA's strategy called for interfering with the powerful TALLINN acquisition radar deployed throughout the USSR to alert the SA-2 and interceptor units. TALLINN was NATO name referring to the P-14 a 2D VHF radar developed and operated by the former Soviet Union. The P-14, the first high power VHF radar to be developed by the Soviet Union, entered service in 1959 following the successful completion of the radar's test program.

The TALLINN used a single antenna accomplished both transmission and reception. The antenna was a large open-frame truncated parabolic antenna capable of modulating its frequency around four pre-set frequencies to counteract active interference.

To defeat the Soviet radar systems, the OXCART A-12 incorporated the first meaningful stealth protection against high-frequency microwave signals like those used by the SA-2 tracking radar. By contrast, the TALLINN relied on low-frequency VHF signals and could override the A-12's stealth protection.

These observations made it clear to the CIA that it needed to learn a great deal about the characteristics of the SA-2 and TALLINN radars. The CIA needed to know their transmitted power accurately, their antenna patterns to determine their coverage, their receiver sensitivity, and their operational reliability. For OXCART route planning over the USSR, the CIA also needed to know the TALLINN radar locations. The CIA faced enormous challenges to determine if the A-12 could safely overfly the Soviet Union.

Today, the CIA could easily locate the TALLINN radars using space-based ELINT satellites, but not so in the 1960s for lack of such satellites and technology. The CIA received a break when it noticed a wide variety of radar signals regularly reflected by Soviet ballistic missiles during a test flight. The CIS recorded these signals at the DDST intercept site in Iran on the southern shore of the Caspian Sea.

The CIA estimated it is detecting reflections of the powerful TALLINN signals from the moon using a 60-foot antenna that the DDST employed in New Jersey. The proper geometry for intercept occurred only rarely. The transmitter, the moon, and receiver aligned in a unique way that permitted receipt of the moon reflected signal at the CIA site. However, this stringent alignment requirement allowed an individual determination of the radar's location for each intercept. As the moon's relative motion eventually highlighted every possible site in the USSR, establishing the site of every TALLINN in this way.

To measure the transmitted power and antenna pattern of the TALLINN radars, an effort known as the Quality DDST mounted an ELINT program with significant air force support. CIA engineers used carefully calibrated instruments flown in the Berlin corridor on routine supply missions and along the Soviet border on ferret flights to make laboratory-quality measurements.

To design appropriate jammers, the CIA also needed to know how sensitive the TALLINN receivers were. At this point, the CIA team created project Palladium, one of the most delicate and fruitful intelligence programs.

Palladium's technique called for moving in close to one of the TALLINN sites along the border to allow special equipment to make copies of their outgoing signals. The CIA's equipment held the captured signal briefly in delay lines and retransmitted the stored signals back to the TALLINN with varying signal strength.

The National Security Agency helped the CIA monitor tracking reports from the radar site, noting the Soviet radar response to Palladium operators changing the power and range (delay time) of the phantom targets they created. Palladium conducted a hearing test on the TALLINN radar systems and soon acquired an accurate measurement of their sensitivity. As far as the CIA knows, the Soviet never understood what the Palladium team did. The CIA shared its data on the acquisition radars with Strategic Air Command and the Navy to use in their SIOP planning.

To this point, the quality ELINT program at CIA was also making a detailed measurement of the SA-2 tracking and guidance signals using cargo flights to and from Berlin. Nonetheless, the CIA felt that it needed more direct access to SA-2 equipment. They accomplished this during a commando raid on a SA-2 site in Cuba that captured critical components and manuals for DDST designers. In an astonishing coincidence, one of the cargo vessels owned by John McCone was sailing along the Cuban coast just as this firefight began. It suffered considerable damage, and the DCI landed on Hick Helms and Dr. Wheelon with great vigor.

Poteat's team decided that it needed to handicap both the TALLINN and SA-2 radars to ensure the safe passage of OXCART over the USSR.

The first approach used existing military ECM units. In a private meeting with the Joint Chiefs, John McCone, and Dr. Wheelon requested the use of the ALQ-49 and ALQ 51 jammers then in production. However, the Joint Chiefs of Staff declined for fear of the US planes falling into enemy hands.

DDST immediately began the development of special-purpose jammers to protect our reconnaissance aircraft, an easier task than that faced by SAC because the OXCART A-12 flew faster and higher, and had a much-reduced radar cross-section.

One of the major contributors to Information Warfare, i.e., stealth, countermeasures, and ELINT was CIA electrical engineer Gene Poteat, program manager for payloads and ECM for the CIA's U-2 and A-12 reconnaissance planes developed at Area 51. Poteat worked in DPD at CIA that later became DDS&T, under Dr. Wheelon. He collaborated with John Parangosky and Dick Bissell from 1960 to 1975.

Poteat's engineering approach to stealth was creating an airplane that resulted in a small unnoticeable blip on enemy radar screens. He accomplished this by shaping the plane with razor-sharp edges, or chines, by tilting the rudders inboard to reduce radar reflections, and by using as much composite radar-absorbing material as practical. Our question was how small a radar target was small enough? That depended on how good the Soviet air defense radars. This project raised more intelligence questions than answers about the Soviet air defense radars.

The Intelligence Community (IC) had no hard information about the transmitter power of Soviet radars, their receiver sensitivity, the spatial coverage of their beams, or even how widespread their deployment. The CIA's Clandestine Service did not have a single officer assigned to the Soviet Union because the US Ambassador in Moscow forbid it.

Then, ELINT could not provide answers to such hard questions. Further, few in the ELINT community knew anything about the A-12 Project OXCART, and fewer still knew anything about the stealth aspects of the program. It seemed to come down to making the best intelligence estimate possible regarding Soviet radar capabilities for dealing with a high and fast airplane with a small radar cross-section. In the words of other intelligence veterans, "Estimating is what you do when you do not know and cannot find out."

To understand why ELINT and the information estimating process had so little to offer, and how the intelligence agencies regarded both, one needs to take a closer look at both in the early years of the Cold War. The first decade or so after the National Security Act of 1947 created the CIA and the National Security Agency (NSA), along came the three disciplines are known today as SIGINT. These are:

Communications intelligence (COMINT), derived from NSA's intercept, decryption, and analysis of foreign communications.

ELINT, based on the interception and analysis of signals, other than communications. (Such as radar and other signals associated with weapon systems, and carried out by practically every element within the IC.)

Telemetry Intelligence (TELINT), usually from the collection and analysis of telemetry from missiles in flight, mostly by the CIA.

These three disciplines provided most intelligence available at the time to intelligence analysts in conjunction with aerial photography coming mainly from the U-2s flying over the Soviet Union. The analysts also drew on information from clandestine or open sources. They included their views, biases, and guesses producing National Intelligence Estimates (NIEs), the intelligence "bibles" on Soviet strategic and tactical technologies and capabilities, and on Soviet intentions.

However, three problems cropped up with NIEs. First, no product is ever better than its sources and is often too meager. Second, NIEs often proved dangerously wrong on crucial strategic issues. For example, just before the Cuban missile crisis of 1962, an NIE concluded the USSR not the place for strategic weapons in Cuba, this occurring even with evidence of some missiles there already. Third, the Palladium team often lacked sufficient information available to produce even a guess, much less an estimate, on such esoteric topics as the ability of a radar system to detect stealthy aircraft. When possible, the CIA considered COMINT and photography the most credible sources of intelligence, and they provided the bulk of the NIE contributions.

No Regard for ELINT

ELINT's contribution was virtually nil, and intelligence analysts considered it next to useless. One prominent CIA operations officer said the Clandestine Service thought ELINT a five-letter cuss word. He viewed ELINT as worthless and felt the CIA should rely only on agents for valuable information.

ELINT was mostly a passive, rudimentary means of collection. It involved getting a radio receiver and recorder within line of sight of the Soviet radars or other sources of relevant non-communications signals. From radio direction finding and the recordings, one could well determine the radar's location, and the signal's overall frequency, pulse rate, and pulse width. From these signal parameters, an analyst could then estimate the radar's performance, however, not with any great accuracy or certainty.

The challenge was finding a way to intercept these radar signals beyond the line of sight or the horizon. Equally, challenging was finding a way to intercept those radar signals within our line-of-sight, the ones the Soviets, who understood radiation security, simply kept off the air. The object of all this ELINT collection by various IC elements contributed to the Department of Defense's Electronic Order of Battle (EOB), a publication listing the locations of the different radars or signal sources for some consumers. The EOB was rather incomplete, and thus unreliable because the Soviets kept most of their radars well out of sight of any ELINT collection assets.

This was the scene at the end of 1959 when Poteat was a new engineer assigned to the CIA's ELINT Staff Office (ESO) in the Office of Scientific Intelligence (OSI). (Gene Poteat spent the bulk of his career in the directorate of science and technology.)

The CIA cleared Poteat into the A-12 Project OXCART project and the stealth aspect. Early on, he met a group of OS1 analysts discussing a newly intercepted signal, apparently picked up by an ELTNT site in Berlin. The analysts had sketched the signal's characteristics on a blackboard. Poteat suggested that it probably was a missile guidance signal because of its similarities to guidance signals. Poteat was working with earlier at Bell Telephone Laboratories in New Jersey and Cape Canaveral as a missile guidance development engineer.

The mystery signal indeed was the long-sought SA-2 GUIDELINE SAM guidance signal. Additional ELINT intercepts over the ensuing years revealed enough about the signal to build electronic jammers able to counter the SA-2. Poteat learned later of the Soviets easily acquiring US patent information on which they based the SA-2's FAN SONG radar. The Soviets enjoyed easy access to US technology, while the CIA had hardly any access to theirs.

It mainly concerned the A-12 Project OXCART mission planners them not knowing how widespread the Soviet early warning radars and their locations. It seemed impossible, however, to determine the number, exact location, or any other technical information on those radars. Poteat recalled an occasion at Cape Canaveral in the 1950s, when they picked up the signal from a ground-based radar located 1,000 miles beyond our horizon. The signal reflected off a Thor IRBM during a test flight.

This suggested the CIA was using this phenomenon, later called bistatic intercept, to intercept Soviet high-powered radars well over the horizon. He needed to do so by pointing the ELINT antennas at the Soviet ballistic missiles during their flight-testing. He needed to use the missile's radio beacon for pointing or programming the ELINT antennas to follow the missile's predicted trajectory. Previously, the standard practice pointed the ELINT antennas at the horizon in the direction of the target radars, resulting in no signal interception.

George M. the head of the ESO thought the idea worth pursuing. He suggested Poteat run the idea by a couple of highly regarded ELINT experts from private industry to fine-tune the concept before continuing. Consequently, Poteat went to California to discuss the idea with Dr. William Perry of Sylvania's Electronic Defense Laboratories in Mountain View and with Dr. Albert "Bud" Wheelon of TRW in Los Angeles, both of whom offered technical and moral support. They lacked computers in those days, so the feasibility studies and engineering calculations involved solving spherical trigonometry equations using slide rules, tables of logarithms, and hand-cranked mechanical calculators.

George's approach paid dividends with an unusually swift funding approval and the installation of a finished system. Poteat named it MELODY after one of his favorite-sounding words. He to operate MELODY at CIA's ELINT and COMINT site on the shores of the Caspian Sea in northern Iran.

Over the ensuing years, MELODY produced bistatic intercepts of virtually all the ground-based Soviet missile tracking radars at a test range 1,000 miles away.

The fixed location of MELODY and the limited trajectories of the Soviet missiles tracked. However, this still failed to provide the A-12 Project OXCART planners the needed locations of all the air defense radars throughout the Soviet Union.

A new challenge occurred with the discovery of a new Soviet air defense early warning radar, the TALLINN. The TALLINN began appearing about this time, which, if widely deployed, seemed to improve the Soviets' air defenses significantly. The TALLINN quickly became the nemesis of the A-12 Project OXCART planners. MELODY'S success with the high-powered, missile tracking radars led to the idea of using the moon as a distant bistatic reflector to intercept and locate the Soviet TALLINN radars emplaced throughout the Soviet Union.

At the same time, the Lincoln Laboratory, America's premier radar-development house engaged in a "radar astronomy race" with its Soviet counterparts. They competed to see which side first detected and characterized the moon's surface using radar. Lincoln won, handily. Poteat visited Dr. John Evans at Lincoln Labs and discussed the moon radar results and the bistatic ELINT idea.

The CIA drew on the Lincoln Labs' understanding of the moon as a reflector of radar signals. The CIA attached sensitive ELINT receivers, tuned to the TALLINN frequency, to the 60-foot RCA radar antenna just off the New Jersey Turnpike near Moorestown with the antenna pointed at the moon. The ELINT receivers, optimized for the effects of the moon as a reflector, used the Lincoln Labs' "matched filter" techniques. Over time, as the Earth and moon revolved and rotated, all the Soviet TALLINNs came into view one at a time, enabling the CIA/NSA team to plot their precise geographic locations.

Many TALLINNs found, and the incredibly complete radar coverage of the Soviet Union was not good news for the A-12 Project OXCART Office. Nor was it good news for the US Air Force Strategic Air Command (SAC), which plotted wartime bomber penetration routes.

Lockheed made its estimates of the A-12 Project OXCART'S vulnerability to Soviet radars, which some felt excessively optimistic. Although some earlier efforts quantified the older Soviet radars' capabilities by measuring their power and patterns, they obtained only limited results.

Now assigned to the A-12 Project OXCART office, Poteat asked for, and received the job of trying to get the hard engineering data needed on the threat radars to put the vulnerability issue to rest. Taking the first step, they asked OS1 to set up an exclusive "Vulnerability Analysis Group" to work with the expected hard engineering data. The group worked closely with his ELINT collectors, advising the Palladium on the data needed and by suggesting collection operation ideas.

A talented team looked at the Soviet air defense radars, particularly the TALLINN. They looked to a lesser degree, at all the other major acquisition, target, and missile tracking and guidance radars. The obvious place to start was where the earlier efforts left off. However, they needed to do so with a system that produced repeatable and unquestionable results. Poteat assembled a small group of engineers known for their innovative nature, their understanding of the Soviet air defense system, and their ability to operate anywhere in the world.

The core group never consisted of more than six people. Poteat and his team ran a C-97 flying ELINT laboratory operating in the Berlin air corridors, giving them line-of-sight access to East German-based Soviet radars.

Meanwhile, Richard Bissell ran a similarly equipped RB-47 reconnaissance aircraft operating around the periphery of the Soviet Union, an air force master sergeant, almost a double for the original model for television's Sergeant Bilko, ran the ground operations. PPMS Activity These projects lead to a series of airborne Power and Pattern Measurement Systems (PPMS) that measured a radar's spatial coverage and its radiated power with extreme precision. The PPMS required upgrading for measuring other important radar signal parameters. These including radio frequency coherence, polarization, and internal and external signal structure details, providing even further insight into a radar's performance vital to the Vulnerability Analysis Group and the designers and builders of electronic jammers.

They needed the precise dimensions of the TALLINN antenna for calculating the antenna gain for use in our radar equations. One US military attach6 got close-in ground photographs of the radar in East Germany. The antenna mounted on a small brick base, and the CIA asked for the dimensions of one of the bricks. It turned out the bricks came from the nearby Pritzwalk Brick Factory. When Poteat asked the Clandestine Service to get us a Pritzwalk brick, he dared not admit it was for an ELINT project. He was happy to give the impression of it hollowed out to conceal something.

Poteat installed the PPMS in a series of air force planes, starting with a C-97 and an RB-47, then C-130 and finally modern RC-135s. PPMS missions flown around the world, along with the periphery of all Communist countries, and in the Berlin air corridors. Of equal importance, these projects led to an incredibly close and comfortable working relationship between the CIA, NSA, and the air force. The CIA's Office of ELINT (OEL) published technical reports on the product of each mission. The CIA distributed the reports throughout the defense and intelligence communities, as well as to the industry's electronic countermeasures designers. These reports led to a flood of requests for more information about both old and new radars, which generated more missions.

One of the early benefits of this accurately measured air defense coverage was it revealing the Soviets' low-altitude coverage far better than our analysts' earlier estimates. SAC quickly changed its SIOP plans for wartime penetration to a much lower and survivable altitude. The projects also answered the analyst's question of whether the TALLINN radar also had a height-finding capability for determining an aircraft's altitude as well as its bearing and range. One of our RI3–47 towing its PPMS antenna a mile behind the plane while over the Sea of Japan, abruptly descended 5,000 feet and then quickly climbed back to cruise altitude.

A nearby US SIGINT station confirmed the Soviets' radar is observing the aircraft's attitude change. The plane's warning receivers confirmed no other radars tracking it at the time. Although under strict orders never to deviate from a constant, nonthreatening flight profile near the Soviet border, the pilot explained his actions to his superior as turbulence avoidance.

The CIA now knew the Soviet air defense radars' power and spatial coverage, however, only half the answer to the A-12 Project OXCART'S stealth—and health. The Palladium team also needed to know the sensitivity of the Soviets' radar receivers and the proficiency of their operators.

Project OXCART had its area on the second floor of CIA headquarters at Langley. Even at CIA headquarters, the sensitive activities of Project OXCART operated on the same compartmentalization and need-to-know protocol as it did at Area 51. If not cleared and read into OXCART, they did not ask or tell. The CIA considered OXCART so sensitive that it conducted its top-secret clearance investigations rather than trust the FBI. Anything OXCART took priority over everything else.

For example, any film arriving for the process with the OXCART identifier went ahead of even the U-2 films for processing.

Another example occurred d with Colonel Slater flying the F-101 from Area 51 to Washington, DC to attend the retirement party of Gen Jack Ledford at the CIA. Slater was suffering both a hangover and a bad case of gout when he landed at Wright Patterson to refuel. There was a long line of planes in line for fueling. Impatient to resume his flight, Slater produced an OXCART identification at which time the ground crews removed the fueling line from a general's plane and pushed it out of the way to make room for Colonel Slater's plane to move him to the head of the line. During Operation BLACK SHIELD, the code words COIN READER provided the same VIP priority and privileges as did OXCART.

The A-12 Project OXCART Office had a stable of top outside scientists to draw on. With their help and suggestions, Poteat came up with a scheme to electronically generate and inject carefully calibrated false targets into the Soviet radars, deceiving them into seeing and tracking a ghost aircraft. The CIA code-named this project PALLADIUM.

They received the radar's signal and fed it into a variable delay line before transmitting the signal back to the radar. By smoothly varying the length of the delay line, they could simulate the false target's range and speed. Knowing the radar's power and coverage from the PPMS projects, Palladium could now simulate an aircraft of any radar cross-section. It did not matter if it was an invisible stealth airplane or one making a large blip on Soviet radar screens. It covered anything between, at any speed and altitude, and fly it along with any path.

Bud Wheelon, now the CIA's new deputy director for Science and Technology, dubbed our project PALLADIUM. Now, the real trick was finding some way of knowing which blips the Soviets saw on their radar screens. The smallest size blip became a measure of the sensitivity of the Soviets' radars and the skill of their operators. Poteat began looking at some possible Soviet reactions that might give us clues whether it saw our aircraft. The clues ranged from monitoring the Soviets' communications to their switching on other radars to acquire and identify the intruder. Richard Bissell suggested the CIA team with the NSA to provide the SIGINT monitoring of Soviet reaction to our ghost aircraft. The easily decrypted link responded in real time, thus providing a real key to several PALLADIUM successes.

Every PALLADIUM operation consisted of a CIA team with its ghost aircraft system, an NSA team with its special COMINT and decryption equipment, and a military operational support team. PALLADIUM carried out its covert operations against a variety of Soviet radars around the world. They included ground bases, naval ships, and submarines, the submarine antenna installations as the trickier.

The coordination of such operations often turned into a nightmare. For example, one winter, when heavy snows closed all airports in northern Japan, Jack W. spent more than three weeks transporting his massive PALLADIUM fan by train.

The small rail tunnels forced him to spend about three weeks in northern Japan, in the winter. To haul his van of PALLADIUM equipment, he used trains, trucking, even sledding to get it over the mountains and back on another train on the other side. Once operational, Jack flew his black ghost in and out of the Soviet air defenses.

When the Soviets moved into Cuba with their missiles and associated air defense radars, they installed many of them near the coast. Thus, the Soviet Union presented an excellent opportunity to measure the system sensitivity of the SA-2 missile radar near home rather than in the Soviet bloc.

One particularly significant operation conducted during the Cuban missile crisis had the PALLADIUM system mounted on a destroyer out of Key West. The ship lay well off the Cuban coast, just out of sight of the Soviet radars near Havana, however, with our PALLADIUM transmitting antenna just breaking the horizon.

The false aircraft appeared as a US fighter plane out of Key West about to overfly Cuba.

A Navy submarine slipped in close to Havana Bay with the intent to surface just long enough to release a timed series of balloon-borne metalized spheres of different sizes. The idea was for the early warning radar to track the electronic aircraft and then for the submarine to surface and release the "calibrated" spheres up into the path of the oncoming false aircraft. It took a bit of coordination and timing to keep the destroyer, submarine, and fake aircraft all in line between the Havana radar and Key West.

Poteat hoped for the Soviets tracking and reporting the intruding aircraft and then turning on their SA-2 target tracking radar in preparation for firing their missiles. He hoped for them reporting their seeing the other strange targets, or spheres, as well. The smallest of the metallic spheres reported seen by the SA-2 radar operators correspond to the size or smallest radar cross-section for detecting and tracking an aircraft.

Poteat obtained the answers sought, however, not without some excitement and entertainment. Cuban fighter planes had fired on a Liberian freighter the day before, although the ship's Liberian flag, which is easily mistaken for the American flag, was quite visible. The attack led Poteat to expect the Cubans and Soviets not hesitating to attack a US-flagged vessel. In the middle of the operation, Cuban fighter planes began circling over the spot where the submarine had surfaced, and another fighter plane gave chase to the ghost. They had no trouble in manipulating the PALLADIUM system controls to keep the ghost aircraft always just ahead of the pursuing Cuban planes. The Cuban pilot radioed back to his controllers that he had the intruding aircraft in sight and was about to make a firing pass to shoot it down. They all had the same idea at the same instant. The technician moved his finger to the switch; Poteat nodded yes, and he turned off the PALLADIUM system.

The submarine's possibly lingering on the surface after releasing his balloon-borne radar targets and unaware of the fighters circling above worried Poteat. Poteat asked the destroyer's captain to broadcast a quick, short message, in the open, to the submarine, telling him to submerge and get out of the bay. The Captain passed the word to transmit the message. A keen sailor responded by hitting the intercom button and shouted down to the radio operator below deck, "Dive! Dive" and then added in response to a question from the radio operator, "No, not us. Tell the submarine to do it."

With the remarkable achievements up to now, the CIA felt it knew at least as much about the Soviets' radar air defenses as they did. The CIA also found their radars excellent, modern, and their operators equally proficient. Poteat finished his special mission in support of the A-12 Project OXCART stealth program.

He gave his collected data, now called Quality ELINT, to the OS1 analysts. The analysts then finished their vulnerability analysis job by concluding the Soviets indeed capable of detecting and tracking the A-12 Project OXCART, which came as no surprise to any of them.

The OS1 analysts put it to Poteat differently. They said that he proved the Earth was round. Their point was when the A-12 Project OXCART came over the horizon, the Soviet air defense radars having the capability of immediately seeing and tracking it. At the same time, PALLADIUM also established realistic stealth radar cross-section goals of the next generation of stealth aircraft to allow the plane to fly with impunity right through the Soviet radar beams. The F-117 stealth fighter was the first aircraft to meet these goals.

Chapter 8 - Targeting the OX

After the unhappy end of U-2 flights over the Soviet Union, US political authorities became understandably cautious about committing themselves to further manned reconnaissance over unfriendly territory. No one expressed any serious intention of using the A-12 over Russia. Save for some unforeseeable emergency; no one felt it was necessary to do so. What then, should they do with this vehicle?

The first interest was in Cuba. By early 1964, project headquarters began planning for the contingency of flights over that island under a program designated SKYLARK. Bill Park's accident at the beginning of July held this program up for a time. Nonetheless, on 5 August Acting, DCI Marshall S. Carter directed that SKYLARK achieve emergency operational readiness by 5 November. At Area 51, the CIA prepared a small detachment to overfly Cuba, though at something less than the A-12' full design capability and goal to operate at Mach 2.8 and 80,000 feet altitude.

Meeting the deadline set by General Carter required the CIA validating the A-12's camera performance, qualifying pilots for Mach 2.8 flight, and coordination with arranging supporting elements. Only one of several types of equipment for electronic countermeasures (ECM) was ready by November.

A senior intragovernmental group, including representation from the president's Scientific Advisory Committee, examined the problem of operating over Cuba without the full complement of defensive systems. This panel decided the first few overflights could safely occur without them, however, felt the ECM necessary after that. The delivery schedule of ECM equipment was compatible with this course of action.

After considerable modifications to aircraft, the detachment simulated Cuban missions on training flights and announced a limited emergency SKYLARK capability on the date General Carter had set. With two weeks' notice, the A-12 detachment could carry out a Cuban overflight, though with fewer ready planes and pilots than planned.

During the following weeks, the detachment concentrated on developing SKYLARK into a sustained capability, with five ready pilots and five operational aircraft. They needed to figure out aircraft range and fuel consumption, reach repeatable, reliable operation, and finish pilot training. They prepared a family of SKYLARK missions, and coordinate routes with North American Air Defense, Continental Air Defense, and the Federal Aviation Authority. They did this without hindering the primary task of working up OXCART to full design capability. One may expect the story, however, by remarking that despite all this preparation, the U-2s proved adequate, ending the need to fly an A-12 reconnaissance mission over Cuba. The CIA reserved the A-12 for more critical situations.

Even before Poteat finished his PALLADIUM project, it became apparent that, if the A-12 Project OXCART could not fly stealthily. Nonetheless, it could operate safely, relying on its superior performance to outfly the SA-2 missiles.

The CIA realized the need for a stable of effective electronic countermeasures systems in the future. Poteat's small group had already spun off two other groups. One took on the job of developing electronic jammers and warning receivers for the A-12 Project OXCART and the U-2s still flying, albeit over China rather than the Soviet Union. A second group continued investigations into revolutionary techniques to reduce further the A-12 Project OXCART's radar cross-section to an acceptable level.

The second group came up with some innovative schemes, such as the mounting of electron guns on the A-12 Project OXCART to generate a radar-absorbing electron cloud in front of the plane. The new electronic jammer group, in turn, began to task a new, more responsive generation of ELINT collectors to obtain even more detailed information about the Soviet radar signals.

They developed, modified, and installed one of the U-2 missile warning receivers in an air force fighter plane. The modification became the basis of a later system called WILD WEASEL, used to find, and destroy SA-2 SAM sites in North Vietnam.

The WILD WEASEL became the stuff of great stories and legends. In deeds of derring-do, the pilots hunted down the SA-2 sites. They launched their radar-killing missiles in close and dodged the missiles fired at them during the encounters.

The special OXCART collection projects had taken on a life of their own. Poteat's focus soon shifted to a broader range of other so-called intractable technical problems. His team and he went after the signals others needed, wherever and however, possible.

About this time, satellite photography had shown an enormous new radar deep in the Soviet hinterland, the HEN HOUSE. The analysts "estimated" that it was a phased-array radar with some space surveillance capability. By now, early ELINT satellites in orbit knew the radar's frequency in the VHF band. A second HEN HOUSE under construction a couple of hundred miles inland from Riga in the northeastern Soviet Union stayed well beyond any ELINT receivers line of sight.

Judging from the size of the radar and its probable high power, Poteat felt his exploitation system should pick up its signal even with it not pointing his way. The signal should extend out several hundred miles and scatter forward and over the horizon by way of a phenomenon known as a tropospheric — scatter of radio waves. Poteat found a map showing an island in the Baltic Sea that looked about the right distance from the HEN HOUSE to install a tropospheric-scatter receiver to intercept and continuously monitor the radar.

He undertook extensive negotiations seeking approval for access to the island. Once approved, he installed dual antennas about 50 wavelengths apart that reduced the expected atmospheric fading. He put the receiver on automatic pilot or unattended operation, and then they waited.

The BRIAR PATCH troposcatter system finally picked up the very first HEN HOUSE transmission and every later transmission.

They learned of the radar tracking US satellites from the first orbit, and that the Soviets had an incredibly effective espionage network in place to tip off the HEN HOUSE each time a US intelligence satellite prepared for launch. During any long holds of an impending launch from Vandenberg Air Force Base in California, the HEN HOUSE switched off and came back on the air the instant the satellite lifted off from Vandenberg AFB, California. The launch gave the HEN HOUSE a track of the satellite on its first pass overhead. They also learned the HEN HOUSE tracked planes just as well and as often as satellites.

This idea of monitoring a radar's operations full-time was analogous to the concept of traffic analysis in COMINT. In the HEN HOUSE case, the radar's precise frequency showed its pointing angle, then correlated with the targets under its track.

First Flights

Upon arrival, reassembly of the plane and installation of the J-75 engines began. Lockheed soon found the plane's tank sealing compounds not adhering to the metals, making it necessary to strip the tanks of the faulty sealing compounds and reline them with new materials to stop excessive fuel leaks. Thus, occurred one more unexpected and exasperating delay in the program.

Finally, on 26 April 1962, Aircraft 121 was ready. On that day, per Kelly Johnson's custom, Louis Schalk took it for an unofficial, unannounced, maiden flight lasting some 40 minutes. He flew the plane less than two miles at an altitude of about 20 feet. The following day, Schalk made a 40-minute flight. Some CIA personnel (including Richard Bissell) and Najeeb E. Halaby, head of the Federal Aviation Administration saw Schalk's official first flight, several days later.

As in all maiden flights, the program detected minor problems. However, it took only four more days to ready the plane for its first official flight. Piloted again by Louis Schalk, it took off at 170 knots, with a gross weight of 72,000 pounds, and climbed to 30,000 feet. Top speed was 340 knots, and the flight lasted 59 minutes.

The pilot reported the plane responding well and extremely stable. Kelly Johnson declared it the smoothest official first flight of any aircraft he ever designed or tested.

The A-12 was extremely long, with a slim shape, two enormous jet engines, a long, sharp, projecting nose, and swept back wings appearing far too short to support the fuselage in flight. This described a revolutionary airplane able to fly at three times the speed of sound for more than 3,000 miles without refueling. Toward the end of its flight, when fuel ran low, it could cruise at over 90,000 feet.

The Air Force SR-71 still officially holds the world records to altitude and speed for an air-breathing aircraft. However, with the added weight of the back seater and his life support systems, the SR-71 lost 5,000 feet of altitude enjoyed by the lighter A-12 and speed. Because the existence of the CIA's A-12 stayed classified for 25 years, it could never claim the world speed and altitude it had over the SR-71.

The plane broke the sound barrier on its second official flight, 4 May 1962, reaching Mach 1.1. Again, the pilot and engineers reported only minor problems.

With these flights done, jubilation was the order of the day. The new Director of Central Intelligence, Mr. John McCone, sent a telegram of congratulation to Kelly Johnson. A critical phase triumphantly passed; however, there remained the long, difficult, and sometimes discouraging process of working the aircraft up to full operational performance.

Aircraft No. 122 arrived at the base on 26 June and spent three months in radar testing before engine installations and final assembly. Aircraft No. 123 arrived in August and flew in October. Aircraft No. 124, a two-seated version intended for use in training project pilots came in November. The J58 engines powering it faced delivery delays that prompted pilot training using the smaller J-75's. The trainer flew initially in January 1963. The fifth plane, No. 125, arrived at the area on 17 December.

Radiation Fallout Over Area 51

July 1962 SEDAN, a 104-kiloton thermonuclear explosion, created a crater 320 feet deep and 1,280 feet across on Yucca Flat. The radioactive dust cloud drifted northeast over Groom Pass.

October 1962 Shot BANDICOOT detonated in a subterranean shaft with a yield of 12.5 kilotons. Dynamic venting deposited fallout on the Groom Lake area.

The Cuban Missile Boost

Meanwhile, the A-12 program received a boost from the Cuban missile crisis. U-2s keeping a regular reconnaissance vigil over the island in October discovered the presence of offensive missiles.

Overflights after that became more frequent, however, on 27 October a surface-to-air missile shot down an Agency U-2, flown by Strategic Air Command air force pilot Rudolf Anderson, Jr., on a mission directed by SAC, over Bates, Cuba. This raised the dismaying possibility that continued manned, high-altitude surveillance of Cuba might become out of the question.

The OXCART program suddenly assumed greater significance than ever, and its achievement of operational status became one of the highest national priorities. The gravity of the situation prompted a Lockheed test pilot flying a U-2 against radar sites at Area 51 to evaluate its radar cross-section, the need precipitated by the loss of Rudolf Anderson and his U-2 over Cuba.

Ten engines became available towards the end of January 1963 and powered their first flight on 15 January. Thenceforth, Lockheed fitted all A-12 planes with their intended propulsion system and accelerated flight testing with contractor personnel switching to a three-shift workday.

With each succeeding step into a high Mach regime, new problems presented themselves. The worst of all these difficulties-indeed one of the most formidable in the entire history of the program became known when flight testing moved at speeds between Mach 2.4 and 2.8.

At these speeds, the plane experienced severe roughness that made its operation out of the question. They diagnosed the trouble as the air inlet system, which with its controls admitted air to the engine. At the higher speeds, the uneven flow of air caused the engine not to function properly. Only after a long period of experimentation, often highly frustrating and irritating, was a solution reached. The problem further postponed the day when the CIA could declare the A-12 operationally ready.

The end of 1962 found two A-12 planes engaged in flight tests. The A-12 achieved a speed of Mach

2.16 and altitude of 60,000 feet. Progress stayed slow, however, because of delays in the delivery of engines and shortcomings in the performance of those delivered. One of the two test planes flew with two J-75 engines, and the other with one J-75 and one J58.

It had long since become clear that Pratt & Whitney felt too optimistic in their forecast. The problem of developing the J58 up to OXCART specifications was proving a good deal more recalcitrant than expected. Mr. McCone judged the situation seriously, and on 3 December he wrote to the president of United Aircraft Corporation.

"I have been advised that J58 engine deliveries have been delayed again due to engine control production problems. By the end of the year, it appears we will have barely enough J58 engines to support the flight test program adequately. Furthermore, due to various engine difficulties, we have not yet reached design speed and altitude. Engine thrust and fuel consumption deficiencies at present prevent sustained flight at design conditions which is so necessary to complete development."

Tanker Support

In the spring of 1965, project headquarters began contingency planning to deploy the detachment for possible A-12 overflights in the Far East. The plane was still incapable of reaching its design range, making it infeasible to mount such operations from Area 51. Thus, the CIA found it now necessary to plan for an overseas A-12 operating base in the Far East. At Kadena Air Base, Okinawa, the existing forward Detachment supported only the KC 135 refueling tankers that under the original OXCART concept of operations, operated from the forward overseas bases to provide refueling support of the A-12 at Area 51.

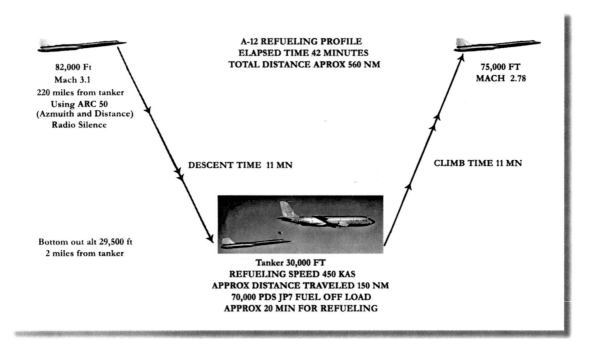

The 903rd Air Refueling Squadron with 23 KC 135 modified aircraft stationed at Beale AFB handled tanker support to both the SR 71 and the A-12. Each plane needed the support of one tanker for each refueling in the ZI, (Zone of Interior), the part of the theater of war outside the theater of operations. The deployment to Kadena, by either plane, needed three air refueling en route with a primary and an air spare tanker supporting each deployment or operational air refueling. During operational periods, mission frequency dictated the tanker support. There were 52 tanker sorties per month needed for A-12 aircrews.

The aircraft had installed in the tail section, an ARC 50 system which provided distance and azimuth to the Tanker, whereas the A 12 had a receiver placed in it to receive the transmitted data. This system was very accurate and necessary as all flights were under radio silence conditions. Quite a feat when considering descending from 70,000 feet at Mach 3 speeds and then leveling off at one mile behind an airborne tanker cruising at approximately 20,000 feet at 600 miles per hour.

An A-12 Article is approaching a C-135Q for refueling. Refueling the A-12 took 42 minutes and 560 nautical miles. Traveling at Mach 3.1 and 82,000 ft, the pilot started his descent 220 miles from the tanker, dropping to 30,000 ft and a speed of 4500 KAS. CIA via TD Barnes collection

An A-12 Article is approaching a C-135Q for refueling. Refueling the A-12 took 42 minutes and 560 nautical miles. Traveling at Mach 3.1 and 82,000 ft, the pilot started his transferred from tanker; it accelerated to 350 knots IAS (its airspeed limit). Max distance between refueling about 2800 nm (3200 statute miles). CIA via TD Barnes Collection

A-12 Trainer - the Titanium Goose approaching to refuel. Note the second cockpit on the trainer

In-Flight Refueling (IFR) Techniques.

The A-12s and the refueling tankers developed and established a most singular, and certainly a first in tactics for air refueling rendezvous and air refueling techniques. The fundamental problem that the CIA and its contractors solved encompassed descending the A-12 from very high cruise altitudes above 85,000 feet, and slowing it from high supersonic speeds up to Mach 3.2 to the tanker. The tanker flew at a comparatively low 30,000 feet air refueling altitude and a speed of Mach.80. To line up directly behind the tanker in position for a hookup, the A-12 had to begin its descent while 240 nautical miles from the rendezvous point. The complete rendezvous and refueling operation from cruise altitude back to cruise altitude encompassed some 700 nautical miles.

The pilot accomplished finding the tanker using the A-12's Honeywell INS navigation system and a discrete UHF radio system that provided range and azimuth information. Additional rendezvous equipment included an airborne beacon and a tactical air navigation system, referred to by the acronym TACAN to provide the A-12 pilot with bearing and distance to the tanker. After the refueling, the A-12 disengaged from the tanker and climbed back up to the cruise Mach speed and continued the mission,

Early on, in 1963, the CIA, wishing to establish air refueling tracks and attendant procedures for operating the A-12 in the Polar areas, had obtained approval to overfly Canada. The intent, of course, was consistent with the initial OXCART concept which envisioned the Sino-Soviet Bloc as the primary area of interest. Subsequently, political interest and emphasis shifted to the China/Southeast Asia area, which required concentrating training routes over the southern United States land mass and some Pacific Ocean areas,

The Vanguard for Deployment

What Weiss did at Area 51 to set up an infrastructure for the people working there, he did likewise for the BLACK SHIELD deployment at Kadena AFB, Okinawa.

Weiss loved a challenge. As an example, before the A-12s deployment to the Far East the CIA had constructed housing meant for the unit's personnel. However, when the CIA received orders for the A-12 deployment, the air force was using these billets in support of Vietnam operations and at a rate four times greater than planned for the 1129th. Weiss decided to make every attempt to find another billeting area paramount to a successful deployment.

Weiss found an abandoned and scheduled for destruction Quonset hut area a short distance from Kadena AFB he decided they could rehabilitate. Colonel Hugh "Slip" Slater stayed doubtful, however, after listening to his plans and concurring that the air force needed our billets he agreed. In a matter of a few months, he restored the Quonsets to better than new condition and added a complete mobile messing facility. A-12 hangers lacked certain features that needed adding just days before the arrival of the first plane. Within a couple of days, Weiss got the significant work underway and completed it on time.

Project LITTLE CREEK

While Weiss was arranging accommodations for the BLACK SHIELD personnel, the CIA Headquarters was doing the same for the missions to follow. The CIA sent Mr. Bill Welsh, as the CIA's communications officer for the Oxcart program, and others assigned to the OXCART program to Kadena Air Base in Okinawa in April of 1964. Welsh arrived at Okinawa and found a three office and one vault building in the Strategic Air Command area in the elephant grass. The air force assigned three blue suiters to him. The CO was Maj Greg Whitney, a fuel specialist T/Sgt Satterfield, and A1c Longleaf, an enlisted man for communications. They were known to the base as Project Little Creek with no explanation of what that was. They were one of three forward staging sites on standby for the deployment of the A-12.

CIA Headquarters decided in late 1965 to make Okinawa the Project launch site. The CIA sent TDY personnel from all over the A-12 project to the Kadena post and built it into a complete project office with full communications, ready rooms, briefing rooms, both remote and local fuel depots. They added wings to each end of the original building and erected a complete antenna field in the elephant grass behind the building.

Welsh was responsible not only for the communications but was also the acting security chief, acting administration chief, acting travel clerk, acting transportation officer, go-for-food gofer, and continuous beer supplier. A team of seven agency communications personnel arrived to run all the lines and install all equipment, not just their Commo gear, but all the phones for the new building, alarm systems, and all other electronic equipment installations.

At any one-time Project LITTLE CREEK had up to 175 TDY personnel on station, all billeting in Quonset huts in a remote area off-base. Sixty days before the arrival of the A-12 articles, they completed all the key construction. The communications center was working 24/7 and had all systems ready. They had seen several dates chosen for the arrival of the A-12s, and they would not know until the last minute the time of arrival.

Two days before the A-12s arrived, Welch received a message from project headquarters saying that he was to be off the island of Okinawa ASAP. Langley believed that since he was known as "Project Little Creek," the Press would inundate them as soon as the A-12s landed at the Kadena Air Base. Within 12 hours, they had emptied their house of all furniture, and appliances. (Furniture belonged to the Okie Station) They had their bags packed, and they were on a Northwest flight to Japan and then Area 51 in Nevada.

The 824th Security Police Squadron personnel assigned to Project Little Creek secured the flight line and hangar area for the SR-71 planes arriving in March 1967 on Habu Hill.

Welch stayed at Area 51 for two years, spent six months back at CIA headquarters, and they joined the Idealist U-2 program for three years in Taiwan working the communications center.

The Federal Aviation Agency at Area 51

Meanwhile, in January 1962, the CIA, air force, and the Federal Aviation Agency entered an agreement that expanded the restricted airspace above Groom to 22 by 20 nautical miles. The lakebed now lay at the center of a 440-square-mile box at the heart of the Nellis Air Force Range. They eventually restricted the airspace continuously at all altitudes.

The CIA cleared individual FAA air traffic controllers for the OXCART project; their function to ensure that unauthorized aircraft did not violate the order.

Area 51 a Power Projection Platform

Historically, Area 51 has always been a power projection or force projection facility. The facility had the capability of applying all or some of its elements of national power. Area 51 influenced political, economic, informational, and military on an international scale. Area 51 could and did rapidly and effectively deploy and sustain forces in and from multiple dispersed locations. Area 51 responded to crises, to contribute to deterrence, and to enhance regional stability.

The term Power projection or force projection as used in military and political science refers to the capacity to apply informational, or military force in and from multiple dispersed locations to respond to crises, to contribute to deterrence, and to enhance regional stability. The Central Intelligence Agency's U-2 Project AQUATONE fit this description from the get-go and was now doing the same with Project OXCART's Operation BLACK SHIELD. Operating out of Kadena, Okinawa, the CIA intended to extend and exert its Area 51 soft power projections in technology and operations to global communications and aerial surveillance starting with the Far East and Southeast Asia. In doing so, the CIA was doing what the

United States' European allies had done for centuries.

The ability of the US Navy, the British Royal Navy, and the French navy to deploy scores of ships for extended periods of time away from home were outstanding projection abilities. The British Empire used its industrial-technological power during the 18th century with the British expeditionary force that extended into the 19th Century with the maritime strength of the Royal Navy defending its interests 6,000 miles from the fleet's home port. The era of British gunboat diplomacy that this inaugurated became an exercise in Western power projection used in defense of its commercial and political interests.

France did the same with its Anglo-French expeditionary force sent to shore up the Ottoman Empire against Russian aggression during the Crimean War in 1853, another example of a planned expeditionary power projection campaign using modern technology such as steam-powered warships and telegraph communications.

The difference between the Central Intelligence Agency power projection and the others, including Russia, Japan was that they exerted military power projection, and the CIA exerted its advanced technology and intelligence gathering as power projection to support the military of the US and its allies. One might say that the CIA intended to do with the U-2 and the A-12 assets what Lt Gen Sir Robert Napier of the Bombay army did when his expeditionary force merged logistics and technology with military intelligence to estimate the required size of his enemy and the difficulties of traversing the inhospitable terrain to conduct warfare.

During both the CIA's U-2 Project AQUATONE and the A-12 Operation BLACK SHIELD, Area 51 became a crucial element of power in international relations. The CIA and the 1129th SAS directed and rotated its military forces outside the limited bounds of its territory, showing a worldwide reach. Excellent examples are Project OXCART and Operation BLACK SHIELD.

The CIA and Air Force's 1129th Special Activities Squadron flew 2850 missions with the top-secret, Mach 3+, high-flying A-12 Blackbird reconnaissance plane.

The CIA at Groom Lake rotated three of its A-12, planes, and CIA pilots from Groom Lake to Kadena, Okinawa. From Kadena, the CIA flew 26 reconnaissance flights over North Vietnam to find SAM sites, and three over North Korea during the USS Pueblo incident. After Operation BLACK SHIELD, the Project OXCART planes and pilots returned to Area 51. The CIA declassified the power projection role Area 51 played in Southeast Asia. The CIA codenamed it Operation BLACK SHIELD.

Initiating Project BLACK SHIELD

On 18 March 1965, John A. McCone, who replaced Allen Dulles as DCI on November 29, 1961, and Secretaries Robert McNamara and Cyrus R. Vance discussed the extent that hazards to U-2 and drone reconnaissance of Communist. China had increased. They agreed to at once take all preparatory steps necessary to run the OXCART out of Okinawa to fly over Communist China. While they decided to move forward with all construction and related arrangements, they did not authorize the deployment of the OXCART to Okinawa or make the decision necessary for flying the OXCART over Communist China,

However, the decision did authorize the preparatory steps and the expenditure of the needed funds The Deputy Director Central Intelligence, General Marshall S. Carter, transmitted Mr. McCone's memorandum to the Deputy Director, Science & Technology, Dr. Albert Wheelon, for action.

Putting the CIA Pilots Back in Their Blue Suits

On 22 March 1965, Lt Gen Marshall S. Carter, deputy director of the Central Intelligence Agency sent the deputy director for Science and Technology a memorandum about Aerial Reconnaissance of Communist China. In it, he informed the DDS&T of his discussing with Colonel Ledford a phone call he had with Colonel Geary regarding a briefing on this subject for Secretary Vance and his discussing the matter with

Dt. McMillan. Geary informed him of fuel for the OXCART already in place on Okinawa. Communications were operational, and the CIA had an operations building and an adequate runway but had not found an available hangar or any other means of putting the CIA's A-12s undercover. Nonetheless, Geary felt the CIA could meet the 1 October readiness date.

Of significance was Geary's statement concerning putting the CIA's A-12 drivers (pilots) back into the blue suits, but allowing them to keep the same rate of compensation. They both agreed that the pilots could not accept any such rigged-up deal like this even if it were possible. Geary said that going into this whole operation on a temporary basis and looked upon as only the first move and that we should make the facility permanent. He stated that they were talking about several million dollars of facilities. Geary also said it was a fallacy to think the pilots have more protection if they were in uniform. Regardless, the air force judge advocate general informed Geary that it was legally impossible to reimburse the pilots over and above regular military pay if they returned to their blue suit status. It took two pages for this Judge to say an unequivocal "no, it cannot be done."

The gist of Geary's memorandum and the related phone calls was the President having the final word on all their issues. The decision rested with the president about returning the sheep-dipped pilots to wearing their air force uniform. It was up to the president to decide on letting the CIA fly the OXCART operationally over Communist China. It was up to him whether the aircraft and pilots would be military with military markings and military pilots, or civilians with deniable characteristics of aircraft and pilot. Given these issues, they all felt it up to the president to decide whether the CIA or the Strategic Air Command would conduct the OXCART operations

There was an anomaly in all of this that beggared analysis. Vance and McNamara became more adamant than the State Department in opposing the use of military U-2s and US military pilots over Communist China. However, at the same time, they now proposed utilizing military planes and military pilots in a much less deniable and much more flagrant violation of normal over flight procedures. Nonetheless, the president and only the president could decide as to whether to use US military pilots with air force markings, or civilian pilots, with no markings. This obviously was a decision of the greatest importance involving national policy at the highest level. The essential point was deniability. The CIA took the position that only it could properly conduct the operation if it was carried out on a deniable basis. The CIA did not believe the Chinese would believe a "blank stare-· never-heard-of-it" basis since it was plain that the Chinese were aware of the overflights. Thus, it was the CIA's position that only the CIA could conduct the operation in this manner.

On. 22 March, BG Jack Ledford, the CIA Assistant Director / Special Activities, AD/SA, briefed Secretary Vance on an operational concept for a Far East OXCART operation codenamed Project BLACK SHIELD. Secretary Vance that the Director/National Reconnaissance Office, Dr. Brockway McMillan, had ordered him to give the facilities and necessary support to implement the plan. A joint CIA/air force team sent to Kadena surveyed the facilities and construction requirements. The team rated the special fuel storage and communications facility as satisfactory the runway adequate, and the venue centrally located relevant to potential target areas. The combination made Kadena the logical choice location for deploying the OXCART operating detachment of Area 51,

Secretary Vance assured the necessary support capability would be available by early fall of 1965 by his making available $3.7 million. OXCART planning for Operation BLACK SHIELD called for staging three A-12s TDY (temporary duty) to Okinawa for 60-day periods twice a year with about 225 personnel involved. A second phase would prove the BLACK SHIELD capability as a permanent detachment,

On 3 June 1965, the secretary of defense noted to the National Reconnaissance Office director that the Soviets were deploying SAM's around Hanoi, and queried the practicability of substituting OXCART planes for U-2s and drones. On 8 June 1965, the National Reconnaissance Office director found the controlling factor in the use of OXCART as the question of performance, operational, readiness, and aircraft reliability, and secondly, the question of vulnerability. He told the secretary of a developing program for making valid determinations of the fuel consumption, range, and other operating parameters of the airplane in its final configuration,

The National Reconnaissance Office director said that he would report an analysis of vulnerability and aircraft performance verification of operational aircraft by 1 July,

At Area 51, with a possible deployment overseas in the fall, the CIA's 1129th SAS Air Force Detachment conducted a comprehensive reliability program to validate standards for a complete system. The CIA set the mission parameters at Mach 3.05, 2300 NM range, and penetration altitudes of 76,000 feet with a demonstrated capability for three, aerial refueling part of the validation process,

The post-modification test and evaluation program saw a considerable improvement in aircraft performance and the operation of ancillary systems. However, while the inlet, camera, hydraulic, navigation and flight control systems all proved acceptable reliability, the sustained time flew at the high Mach, the high-temperature environment had surfaced new problems. The worst among them was the electrical wiring system,

The A-12 needed electrical wiring connectors and components able to withstand temperatures over 800° F along with structural flexing, shock, and vibration. Repeated malfunctions occurred in the inlet control, communications equipment, ECM systems, and cockpit instrumentation, all attributable to wiring failures and careless maintenance. The "now inherent" fuel tank sealant problems continued with no known compound that could withstand the range of temperatures and structural flexing. Worse yet, transducer and indicator failures, air conditioning failures, oscillations in, the inlet system and a host of different bugs continued to plague air operations. The situation needed frequent inspections to assure the durability of the engine. Improving the durability of engine components, such as the combustion section, and the accuracy of the engine control systems still needed flight and ground test development,

Indications of lessening interest in the program and a lack of aggressive interpretation of flight test results were now causing a mounting concern about meeting the BLACK SHIELD readiness date schedule. The CIA project management felt the program needed prompt corrective action on the part of Lockheed. There was no doubt that the quality of maintenance needed drastic improvement. The responsibility lay squarely upon Lockheed for delivering an aircraft system with acceptable reliability to meet an operational commitment. Project headquarters tapped Mr. John Parangosky, deputy for Technology, Office of Special Affairs to ensure the CIA maintenance supervisors at Area 51 observed Lockheed procedures and progress in solving the new problems.

John Parangosky, a Central Intelligence Agency officer since 1948 was born in Shenandoah, Pennsylvania in December 1919, the son of a naturalized Lithuanian father and an American mother. He served in the US Army Air Force and Counterintelligence Corps in the United States and Asia during and after World War II. After the war, he attended the University of Pennsylvania and Columbia Law Schools before joining the CIA, where he first served as a reports officer for seven years in Trieste, Italy. From there, he served in an outpost on the Cold War frontier inside the Communist East Bloc,

In the mid-1950s, Parangosky became part of the management team of the CIA's U-2 under project manager Richard M. "Dick" Bissell. In 1956, he switched to the A-12 OXCART project as the program executive officer and was now the program manager, a position that he had held since the first flight test of the A-12 in 1962.

Mr. Parangosky met with Mr. Kelly Johnson at the Lockheed plant on 3 August 1965 to ensure the program making the BLACK SHIELD and OXCART commitments. The meeting concluded with Mr. Johnson agreeing that he should go to Area 51 on a full-time basis to get the job done expeditiously. Mr. Daniel Haughton, President of Lockheed, concurred with the proposal and offered the full support of the corporation regarding senior people or any other help to achieve BLACK SHIELD readiness. Mr. Johnson arrived at Area 51 for full-time duty the next day. Also, he augmented the Lockheed contingent at Area 51 with senior inspectors, electrical technicians, and manufacturing people,

Having top-level supervisors with boots on the ground at Area 51 paid off with an instant improvement in maintenance crew performance. Inspection procedures improved, and Lockheed management at Area 51 tightened up the contractor reorganization to get project BLACK SHIELD back on schedule. The CIA identified four primary BLACK SHIELD aircraft that went into a flight program upon Lockheed completing the modifications that validated BLACK SHIELD operational profile sorties. Flight test aircraft proved the following performance milestones during 1965.

Maximum Speed	Mach 3, 29.
Maximum Altitude	90,000 feet.
Maximum Sustained Time.	
At, or above Mach 3.0,	1:17 hours.
At or above Mach 3.2	1:14 hours.

During BLACK SHIELD validation flights, Detachment planes, flown by Detachment pilots, recorded a maximum endurance flight of 6:20 hours duration and another on which they logged 3:50 hours of Mach 3.0 or above time,

On 14 August, an A-12 flew an endurance flight from Area 51 that covered 6500 nautical miles in 5:27 hours that included air refueling time. The flight took it to Orlando, Florida, and back to Area 51, thence to Kansas City and return to Area 51. The flight, simulated, exactly as planned, an operational mission with two air refueling and three cruise legs,

DUTCH NUMBERS

Project Pilots	Lockheed:
Frank Murray - Dutch 20	Bill Park - Dutch 50
Ken Collins - Dutch 21	Bob Gilliland - Dutch 51
Jack Layton - Dutch 27	Jim Eastham - Dutch 52
Denny Sullivan - Dutch 23	Art Peterson - Dutch ?
Walt Ray - Dutch 45	Darryl Greenameyer Dutch ?
Mele Vojvodich - Dutch 30	Bill Weaver -Dutch 64
Jack Weeks - Dutch 29.	Joe Rogers - Dutch 68
Bill Skliar - Dutch 69	

1129th Staff
Robert Holbury - Dutch 10
Slip Slater-Dutch 11
Burgie Burgeson-Dutch 12
Jim Anderson-Dutch 15
Ray Haupt-Dutch ?

The SR's flying from Edwards AFB used sixty series Dutch numbers. Except for Layton who used Dutch 72 while he flew the YF's. 72 was the reverse of his old A-12 callsign of Dutch 27.

Project SCOTCH MIST

Between 4 and 15 August 1965, the CIA deployed an A-12 plane supported by an Area 51-based F-101 and a C-130 to McCoy AFB, Florida, a joint-use air force/civil airport near Orlando for a series of climatic tests. The purpose was to figure out vehicle performance in areas of high humidity such as they might experience in Okinawa. The goal was environmental testing to evaluate cockpit fogging, hot air de-icing, and windshield rain removal,

Thus far, the A-12 had flown only out of Area 51 with an altitude of 4,462 feet. The A-12 flown by

Walter "Walt" Ray (Dutch 45) landed at McCoy at an only 82 feet above sea level and during a heavy rain.

Thus far, the project had tested the rain clear system only by flying through local showers. Now, that the plane was landing in meteorological conditions like those expected at Kadena, Okinawa, Walt Ray decided to test the system. With the runway at McCoy in sight, the activated the switch to release a chemical water wetting agent to nozzles located along the Barracuda blade situated at the apex of the windscreen to dissipate the raindrops. To his surprise, the nozzles sprayed a white, milk-like liquid over his windscreen that blocked his vision of the approaching runway. He could not see a thing other than the runway on each side of the plane. Project A-12 pilot Dutch 23, Dennis Sullivan was at the mobile and had to remind Ray to deploy the chute. Ray learned the hard way that the fluid in the rain clear system needed regular changing. Otherwise, it coagulated from the accumulated heat generated during the flight.

The rain clear incident was mild when compared to what happened next. Ray had rolled out for a takeoff, and once in place, he eased back on the throttle. To his and everyone else's surprise, the two J58 engines stopped. There he sat on the end of the runway with no way to restart his engines. Restarting the engines on the ground required a start cart using two Buick 401 cubic inch Wildcat V8 internal combustion engines that positioned underneath the J58 and connected to it by a vertical drive shaft. The powerful engines spun at 3,200 revolutions per minute until the turbojet could self-sustain.

Unfortunately, no one had thought they would experience the engines stopping and had not brought along a start cart. Hence, the CIA's ultra-secret A-12 plane had to sit in the rain at the end of the runway and exposed to the public while the 1129th at Area 51 rushed a start cart to Florida. Meanwhile, Pratt & Whitney, Lockheed, and United Aircraft Co. rushed fuel control and J58 experts from the Skunk Works in California and Connecticut to decide the problem, which turned out being the T-2 sensor in the engine nacelle needing adjustment.

For the first time, passengers in the civilian planes got to see the A-12 Blackbird. Those on the ground got to experience the deafening sounds of two powerful Buick engines screaming with no exhaust pipes. Following that were the monstrous sounds of the two 34,000 pounds-force turbo-ramjet J58 engines. They saw and heard the most powerful air-breathing aircraft engine ever devised, producing more than 160,000 shaft horsepower as Walt Ray took off on the return flight to Area 51.

The data obtained from the flight testing at McCoy showed that the A-12 descending into moist, warm climates presented no flight safety problems, thus proving the feasibility of operating in such a climate as that of Okinawa.

On 20 November 1965, completion of the BLACK SHIELD validation found the aircraft system performing with acceptable reliability and repeatability. The detachment was manned, equipped, and trained, its two primary camera systems performing according to specification. Now started the shipping of a considerable amount of material by surface to Kadena in support of the deployment. CIA's Werner Weiss had the Kadena facilities mostly completed when he and Colonel Holbury received an implementation schedule calling for an operating capability at Kadena Air Base on 15 January 1966. To strengthen the declaration of operational readiness, he was soon to make, the D/OSA, General Ledford, called the principal contractors to a meeting at headquarters on 23 November 1965. There, he received from Lockheed (aircraft and overall systems) Pratt & Whitney (engines) Minneapolis Honeywell (INS and Flight Control System) Perkin Elmer and Eastman Kodak (Cameras) a written statement that each contractor felt his system was ready for a successful BLACK SHIELD operation.

On 1 December 1965, a proposal sent to the 303 Committee via the National Reconnaissance Office director sought approval of the OXCART Far East deployment. On 2 December 1965, the National Reconnaissance Office director endorsed it to the 303 Committee, who examined the matter the next day. Rather than approve an actual deployment as asked for, it agreed, short of moving plane, to have the CIA's Operation BLACK SHIELD ready to develop and keep a quick reaction deployment capability. Operation BLACK SHIELD would deploy within a 21-day period any time after 1 January 1966,

OXCART 1965

The year 1965 saw the CIA conducting 600 A-12 flights totaling 1000 flying hours. It had sharply increased Mach 3.0 flight time that compared to only 8 hours total Mach 3.0 time acquired before 1965, had accumulated 108 hours in 1965. Detachment sortie effectiveness, where all subsystems performed correctly, and all planned objectives of the sortie were satisfactorily accomplished, rose from a low of 25% in 1964 to 65% in December 1965. The OXCART Program could now conduct reconnaissance of Cuba and deploy to the Far East, simultaneously if needed. A third aircraft loss, however, clouded the year's accomplishments.

There was no Flight Simulator developed for the A-12 Program. The pilots, mission planners, and engineers wrote the Dash One from experience. There was no computer, only the slide rule.

On 28 December 1965, Article 126 crashed immediately after takeoff at Area 51. The pilot, Mele Vojvodich, Dutch 30, ejected safely at an altitude of 150 feet, his right foot touching the runway only seconds after his chute opened. The accident investigation-board found the cause being a flight line electrician reversing the yaw and pitch gyros cable connections of the plane's stability system, which resulted in the loss of control of the plane performing a series of violent yawing and pitching gyrations during liftoff,

The director, Central Intelligence directed the Office of Security to conduct an accident inquiry. The investigation concluded that the crash occurred as the result of unintentional errors on the part of the workers directly involved. The investigation found no sign or evidence of malicious intent, sabotage, or wanton disregard of proper procedures. The accident received no publicity and concluded the need for closer technical supervision of maintenance personnel as corrective action. The accident report placed increased emphasis on formal training courses and higher maintenance standards.

Assignment of responsibility for reconnaissance of targets in the Far East solidified employment concepts and tactical training on which the CIA focused BLACK SHIELD planning. Training for three anti-radar missions, over water rendezvous, single engine refueling skill, subsonic missions, and ECM procedures, received heavy emphasis. The conducting of pilot fatigue studies verified the pilots' capability to perform the long duration flights inherent to subsonic deployment to overseas locations,

The Far East subsonic deployment required over water and air refueling that validated a decisive rendezvous system, making supersonic deployments feasible by materially reducing the staging logistics requirement and pilot fatigue factors.

All-important, the OXCART's basic in-flight and photo flight line tactics had evolved from the U-2's operational, experience. The A-12 speed, altitude, and ECM made the aircraft invulnerable to enemy air defense reactions. However, this required the development of special techniques in planning camera flight lines. This was due to the high-speed characteristics and 86-mile turn radius of the aircraft that required planning solid flight lines over primary target areas. Though requiring a fixed flight line over the primary target, the mission could obtain bonus photography during a coordinated turn,

Writing the Dash 1 for the A-12

The Lockheed Aircraft Company recruited Fred White during his last year as a mechanical engineering major at U. C. Berkeley. Shortly after the first flight of A-12 Article 121 in 1962, Chief flight test engineer Ernie Joiner from the U-2 project approved his transfer to the Advanced Development project group. There, White became a member of Bob Klinger's flight analysis division where he and Roger Christensen directed and produced the flight manuals needed for the operation of the A-12, YF-12A, SR-71 and U-2 planes.

Writing the Dash 1 flight manual for the A-12 was a work in process as had been the case with the U-2. The procedure followed specification MIL-M-7700, but only for guidance. Much of what went into the manual came from experience and events reported. White and the others on his staff received input from the pilots, the flight engineers, mission planners, command staff, medical professionals, and the various contractors maintaining the different systems incorporated in the A-12. No one had ever produced a manual for operation of aircraft designed for sustained operation at supersonic Mach 3 speeds and temperatures never experienced with a manned plane.

The previously published U-2C manuals compiled by Lockheed's Ernie Joiner's people helped set up a protocol. However, the supersonic blackbirds were much more complicated for writing suitable and understandable descriptions. Majors Raymond "Ray" Haupt, Sam Pizzo, and Harold Burgeson were instrumental as were Henry Oda and Bill Corbin from the project side. Robert Klinger and Glenn Fulkerson from ADP also helped. Mr. Jack Maxwell produced the checklist changes overnight for next day aircraft delivery to the project pilots. Lockheed's Roger Christensen and Carl Nachtrieb continued working on the flight manuals when White retired from Lockheed,

Chapter 9 - Mission Planning.

The formulation of operational tactics was a continuing effort designed to optimize mission capability consistent with requirements and safety. The mission planning staff developed an A-12 tactical doctrine manual, and the CIA formalized it to govern all A-12 operations. This handbook prescribed in detail A-12 operational planning factors, tactics, and procedures for the employment of the A-12 reconnaissance system.

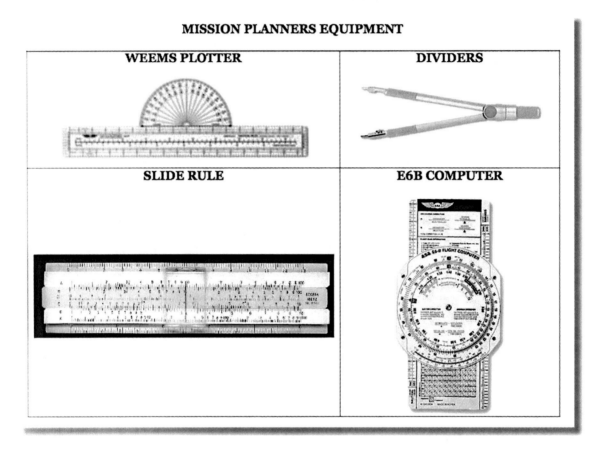

Mission planning was in the primitive state of manual manipulations and computations utilizing data subscribed by contractor specifications. OXCART program mission planners could rely only on the computerized techniques developed in the two prior years to work up a dossier of BLACK SHIELD missions. That changed in November of 1964 when the mission planners had compiled enough graphs, charts, and benchmark data to begin a computer program which eventually afforded an automated flight plan.

This limited capability required the mission planners' eliminating each manual function of planning. From this, emerged in the summer of 1965 a flight plan format that included an automated magnetic variation, and the minimum fuel required at the designated bases. However, it was apparent that mission planning needed a more definitive planning for systems capabilities and vehicle performance. The A-12 validation period made significant progress in automated mission planning that included the following major accomplishments:

(1) Automated plotting capability,
(2) Post mission plot to inertial navigation system tape,

(3) Distance to "go" information.
(4) Extraction of select data in flight plans for NPIC's use in correlating mission "take."
(5) Sun relative bearing with its change rate per minute.
(6) A defense analysis program to give the probability of mission success.
(7) Inertial navigation systems program.

The worldwide capability of the INS system, testing, and validation of all weapons systems and repeatable performance information all came from the validation period ending in November 1965. The mission planners gleaned data from this validation period data the A-12 turn, radius, automated weather data, and fuel curves for the different types of cruise profiles that they incorporated into the computerized flight plan. Additionally, they developed a KC-135 automatic flight plan which became the most advanced system ever afforded a supporting organization. All camera programming data was now included in this program also,

In summary, the computer program could output a complete flight plan depicting all necessary information for planning and employment with a minimum number of specific computer inputs such as departure coordinates and time, target coordinates and destination coordinates. The mission planners tried and proved a customized planning concept with a mission generation countdown of 24 hours,

The computer program proved capable of planning missions at either subsonic or supersonic speed. It did not matter if it received varying power settings, maximum altitude or long-range cruise profiles and varying gross weights. In any case, it automatically computed fuel consumption utilizing the correct parameters.

Normally, the mission planners conducted a mission brief to inform all detachment participants of the peculiar mission needs, Tanker support, camera selection/settings, Commo decisions, mission route, expected weather conditions.

The mission is planning at Area 51 during Project OXCART, and later during Operation BLACK SHIELD consisted of five Officers, three enlisted personnel, and one Airborne Instrumental Laboratories (AIL) Tech. Support individual. These included Sam Pizzo, Bill Corbin, Al Rossetti, Frank Moon, John Clunk, Harold Mills, Ron Mick, Tom Henwood and the AIL Tech Rep, Bill Goodwin, commonly known as "WAG."

Col Sam Pizzo

At Groom Lake, Major Pizzo's planning staff verified the anticipated aircraft performance data and the planning of training sorties that simulated operational missions. This involved map preparation, photographing the map, and preparing the map for cockpit storage. The staff developed a cassette of the photographed map for insertion into the cockpit instrument panel. They determined air refueling locations, coordinated with the tanker squadron, coordinated training routes with cleared personnel within the FAA, briefed pilots before the flight and debriefed pilots after the flight. Moreover, most importantly, verifying that the anticipated performance data by Lockheed performed as projected.

The planning of stateside training routes tasked the planners with limitations such as the restriction of overflying cities of 25,000 or more population due to the Sonic Boom problems. Whenever a training mission created sonic boom complaints, the navy and air force accepted the blame. The FAA's contact for OXCART often called and joked about whom to blame that day.

Mission planning required the planner make up a route map, marking latitudes and longitudes, depicting the desired points to overfly. The planners drew a continuous line of flight from one set of coordinates to the next set. It consumed a significant amount of time to accomplish when considering the distance that the aircraft traveled while traveling at Mach 3 speeds.

They annotated the maps to depict aircraft desired headings, identify visual checkpoints to ensure route adherence, emergency airfields, camera, and electronic equipment on and off locations. They estimated times of arrival to the next checkpoint, emergency airfields, fuel data, and radio frequencies. They photographed the map, cut it to fit the cardboard frames (for placement in the cockpit container).

They placed the film strip in a cassette used to brief the pilots and then put it into the aircraft console when airborne, and viewed by the pilot to maintain course. They also inserted two lines on the map, each equal distance from and paralleled to the flight path. These two lines defined the area where to photo the map before cutting and placing in a metal container located in the cockpit as an emergency map backup if the console display viewed by the pilot to maintain course, failed.

This all occurred before the age of computers. Thus, the planning tools consisted of a handheld set of dividers for measuring distance. Other planning tools included a Weems plastic plotter used to determine the course, a handheld plastic computer (E6B) used to determine time as it related to arrival at checkpoint flight headings, and fuel consumption One of mission planners, Bill Corbin, was a genius with computations and used the slide rule.

The aircraft did not have a navigational radar system. Instead, it utilized the Honeywell INS system (Inertia Navigational System). Visual checkpoints usually seen by crew members flying at 600 mph at 40,000 feet proved no longer valid when flying at 2100 mph at 90,000 feet. Baird Electronics equipped the aircraft with a Baird Electronic viewfinder to provide the pilot with the ability to visually view the ground directly under and to the right or left of the plane. Pilot input on valid checkpoints was always the planners' guidelines in laying out map data.

The OXCART mission planners planned all training routes in close coordination with CIA headquarters. The planning involved much more than routing the A-12. It also included weather conditions along the route, providing F-101 chase planes, and most importantly placing aerial tankers along the route to refuel the A-12.

Beale AFB in California provided KC-135 aerial tankers modified to handle the JP 7 fuel instead of the standard JP 4 fuel used by the air force. The JP 7 jet fuel was a new exotic fuel, a jellylike substance, unlike normal jet fuels, to replace the standard JP4 jet fuel that would explode due to heat generated by the high Mach 3 speed of the A-12. The mission planners worked very close with the aerial refueling squadron's staff to establish air-refueling areas throughout the US. Security concerns remained paramount to prevent exposing the top-secret reconnaissance plane from regular civilian air traffic routes, especially during descent and climb out portions of the refueling.

Refueling the A-12 required the pilot to begin his descent 200 miles before meeting the tanker orbiting at a predetermined rendezvous point. It took a similar distance for the A-12 to resume its mission altitude. The aircraft had installed in the tail section, an ARC 50 system which provided distance and azimuth to the Tanker, whereas the A 12 had a receiver placed in it to receive the transmitted data. This system was very accurate and necessary as all flights flew under radio silence conditions. Quite a feat when considering descending from 70,000 feet at Mach 3 speeds and then leveling off at one mile behind an airborne tanker cruising at approximately 20,000 feet at 600 mph.

A mission often required multiple refueling, each of them planned by the mission planners at Area 51, and Kadena during Operation BLACK SHIELD. The CIA also prepared for each mission, ensuring the presence of CIA security officers along the route to ensure the safety in the event the A-12 required an emergency landing.

At Area 51 and Kadena, the mission planners conducted their briefings on a one to one basis with the A-12 pilot and a member of planning staff in a small room with a projector to carry out the briefing and answer any questions. The maps provided for the mission came from a facility in St. Louis. In an emergency, the pilot could destroy the map by exposing it to water. Operation BLACK SHIELD used disposable maps, however, at Area 51 they did not.

1129th SAS Commander Robert Holbury and Major Sam Pizzo at Area 51

On occasion, at debriefings, our Mission Planners learned of a problem or received a suggestion from the pilots. If appropriate, they passed this information to Lockheed. As an example, planner Al Rossetti changed the mechanism that controlled the viewfinder, this resulting in the change from a panel of switches to that of a wafer switch much easier to operate than turning toggle switches off and on.

The mission planners held a debriefing after every training flight to seek areas needing review and or change. Col Sam Pizzo*, a bombardier during World War II and the Korean War, was now a navigator heading the mission planning at Area 51. He did not deploy to Kadena. Instead, he returned to SAC where the air force selected him to fly onboard Russian aircraft flying in support of Premier Khrushchev's visit to the United States. He made nine trips, eight on civilian versions of the TU-104 and one on the TU-114. The latter returning the Premier to Moscow.

*Colonel Holbury, the first 1129th SAS commander, brought Major Pizza with him to Area 51. At the time, the air force considered an assignment to Area 51 as an unaccompanied tour, meaning they could not bring their dependents. Mary Pizzo challenged the air force about deploying her husband to Las Vegas, Nevada without her and their children. She successfully changed the air force's stance on assigning personnel to Area 51.

Disappearing Maps

An explosive charge placed in the film cartridge area detonated with the pull of the ejection handle. However, there were no means to destroy the maps placed in a metal map case the cockpit.

One day Col Doug Nelson called Maj Sam Pizzo into his office and gave him instructions to go to an agency located in St Louis that made maps for the US Government. Pizzo was to request they develop a standard map for placing in the cockpit map case that the pilot could destroy in flight in the event of an A 12 bailout or emergency over denied territory.

Pizzo went to the agency and showed them what he needed, a map for storage in the metal map case. Two weeks later, he returned to St Louis where the agency workers placed a new map in the cardboard frames used in the A-12. They put the map and frame into in a container and poured a black liquid over the map. It instantly destroyed the map. In-Flight Refueling (IFR) Techniques.

The A-12s and the refueling tankers developed and established a most singular, and certainly a first in tactics for air refueling rendezvous and air refueling techniques. The fundamental problem that the CIA and its contractors solved encompassed descending the A-12 from very high cruise altitudes above 85,000 feet, and slowing it from high supersonic speeds up to Mach 3.2 to the tanker. The tanker flew at a comparatively low 30,000 feet air refueling altitude and a speed of Mach.80. To line up directly behind the tanker in position for a hookup, the A-12 had to begin its descent while 240 nautical miles from the rendezvous point. The complete rendezvous and refueling operation from cruise altitude back to cruise altitude encompassed some 700 nautical miles.

The pilot accomplished finding the tanker using the A-12's Honeywell INS navigation system and a discrete UHF radio system that provided range and azimuth information. Additional rendezvous equipment included an airborne beacon and a tactical air navigation system, referred to by the acronym TACAN to provide the A-12 pilot with bearing and distance to the tanker. After the refueling, the A-12 disengaged from the tanker and climbed back up to the cruise Mach speed and continued the mission,

Early on, in 1963, the CIA, wishing to establish air refueling tracks and attendant procedures for operating the A-12 in the Polar areas, had obtained approval to overfly Canada. The intent, of course, was consistent with the initial OXCART concept which envisioned the Sino-Soviet Bloc as the primary area of interest. Subsequently, political interest and emphasis shifted to the China/Southeast Asia area, which required concentrating training routes over the southern United States land mass and some Pacific Ocean areas,

Pizzo contacted Lockheed and advised them regarding the black liquid, so they could come up with a mechanism to have the liquid flow into the map case upon pulling the ejector handle.

Lockheed contacted Pizzo shortly afterward to invite him to come to the Burbank Plant to view the testing of the system they had devised to destroy the new maps.

A pilot friend flew Pizzo to Burbank in a T-33 jet plane to watch the demonstration. Lockheed placed the map case and the explosive device with the black liquid on a wallboard and started the demo. Pizzo and the pilot stood in the back of the Lockheed personnel who were seated at a long table right in front of the test board when they pulled the switch to initiate the explosion sequence.

The black liquid flew out of the canister into the map case, instantly filling the map case and then spewed forth up and out of the case the black liquid, it heading towards the Lockheed personnel sitting about three feet from the wall. Bodies were jumping and falling in all directions covered with the black liquid. None of it reached Pizzo, who found humor in seeing white-shirted gentlemen from Lockheed scatter trying to get out of the path of that fluid. The liquid did indeed destroy the map.

Training Advances of 1966.

The 1966 training program included an exercise in April to familiarize new mission planners with flight planning preparation and timing. These involved exercises involving film handling and movement procedures that stimulated both SKYLARK and BLACK SHIELD operations. In May 1966, a command post exercise checked out the special procedures for supporting commands in the Far East for BLACK SHIELD to follow. This included forward base exercises in Adana, Turkey, and Kadena, Okinawa that developed new procedures, schedules, and priorities with the Strategic Air Command KC 135 aerial refueling tanker support units.

Liaison and coordination with appropriate US Navy and Air Force command compressed the response time to launch an overflight of Cuba to seven days. From the BLACK SHIELD training exercises, mission planners gained experience that resulted in a revised operations plan that reduced the deployment response time to commence operations at Kadena from 21 to 15 days,

Zl Air Traffic Control Procedures.

In April 1966, project headquarters, the air force, and the Federal Aviation Agency (FAA) conducted a series of meetings to address air traffic control above 60,000 feet within the stateside zone of interior, ZI, outside the Area 51 theater of operations. The increased flying activity by YF 12s and SR 71s out of both Edwards Air Force Base and Palmdale, California at altitudes above 60,000 feet made it necessary to establish procedures for air traffic control at without compromise of project security for the CIA's ultra secret A-12s flying out of Area 51,

The concern was the development and establishment of flight separation criteria for this previously uncontrolled airspace. OXCART personnel needed to establish coded altitude reporting to protect the classified aspect of the operations of the A-12 above 60,000 feet,

The CIA needed mandatory control of airspace in the environs of Area 51 to preserve the security of its Area 51 facility and the flying activity of a supersonic aircraft that publicly did not exist, from a flight test facility that also did not exist. Project OXCART required the establishment of procedures, both unusual and specialized such as Yuletide. Approach Control established in January 1962, ostensibly as the airspace controlling agency for all the Nevada AEC Test Site. It, in fact, did not control the airspace surrounding Area 51 whereas the CIA had an established separate covert approach control that managed and controlled all flight activity there,

Yuletide Special Operations Area

In early 1963, the Yuletide Special Operating Area (SOA) was established to accommodate increased flight test at Area 51; The CIA needed additional airspace to meet the growth in the performance envelope of the A-12. Furthermore, the CIA needed to restrict the use of this designated airspace north of Area 51, making it exclusive for Area 51 aircraft's use between altitudes of 24,000 feet and 60,000 feet and an approximate 150 by 350 nautical miles in an east-west orientation.

The FAA coordinated support aircraft flying along Area 51's arrival/departure corridors using covert procedures that allowed the aircraft to file clearances ostensibly showing them flying to and from Nellis Air Force Base, Nevada when they were in fact coming and going to Area 51,

For the A-12 coming and going from Area 51, the CIA installed a discrete, selective identification feature for secure flight tracking called the Mode X IFF/SIF. In this system, the airborne transponder responded only to specially modified ground radars made compatible with the airborne equipment. The modifications existed only in selected ground radars that few ever knew about that the CIA identified as the Seven Sisters. The Seven Sisters followed all A-12 flying from Area 51 through July 1965 as they flew black without filing a flight clearance. The covert aspect of the CIA's type of flying required extensive coordination. The CIA established specialized procedures with the FAA and NORAD that assured no compromise of A-12 aircraft's existence or performance in the event their radars tracked the A-12,

In July 1965, the A-12 clearance filing procedures began. Visual flight rules (VFR) were regulations under which a pilot operated an aircraft in weather conditions clear enough to allow the pilot to see where the aircraft was going. The procedure for the A-12 ostensibly departing VFR from Edwards AFB was to identify it as an SR 71 and to a fix within the Yuletide. There the supersonic operating area (SOA), picked up an instrument flight rule (IFR), clearance for a route of flight above 60,000 feet. It returned to a fix within the SOA, canceling the VFR flight plan and ostensibly proceeding to Edwards VFR for a landing,

OXCART Pilot Training.

In 1966, the CIA geared individual pilot flight activity to maintaining proficiency and operational readiness. Most training flights included at least one aerial refueling and averaged 45 minutes of Mach 3.0 time. The pilots were checked out in night flying and refueling, Escape, evasion and survival training continued with emphasis on tropical and seacoast areas. The 1129th SAS at Area 51 initiated parasail training as part of the water survival training program,

The Pilot's Emergency Seat Pack

The A-12 pilot's survival pouch contained IAW standard contents with some items added to the medical kit:
Maps
Twelve AA batteries for use in ASR 100 and penlight flashlight
Currency
One pair of socks, white, wool, and nylon
One pair Converse brand tennis shoes.

½ sheet sandpaper, 400 grit, whether dry (used in place of Emery Paper)
Compass, Button
Four each Chloroquine-Primaquine (discontinued tubes of Boric Acid Ointment)
Actifed as a decongestant
Dramamine for motion sickness
Tetracycline, an antibiotic for bacterial infection
Otrivin-aerosol-spray
Cortisporin ointment for skin infections
A-fil, a sunscreen agent
Nupercaine ointment for hemorrhoids
Dilaudid as an opioid pain medication
Lomotil for diarrhea
Other items depended on the mission and venue

The Parasail Boondoggle

Capt Charlie Trapp's Pararescue jumpers had spent much time in the Area 51 pool testing equipment pressure suits, floatation devices and survival procedures with the pilots. Someone suggested parasail training, so Charlie Cravatta, Earl Casto, and Trapp went to Randolph AFB, Texas for ground parasail training and on to Pensacola, Florida for water training with the Navy.

They learned how to run parasail ops.

Charlie Trapp

Pressure suit training resumed following the tragic drowning of Lockheed flight test engineer Ray Torrick in the Pacific Ocean following the destruction of the M-21 mothership during the launching of a D-21 drone out of Area 51.

The A-12 pilots, helo pilot Capt Charlie Trapp, Castro, Bailey, and Col Slip Slater set up the project with the US Coast Guard assigned to Lake Mead. Col Slater had the Coast Guard establish a secure area for the Oxcart project pilots at Groom Lake to practice water survival in their flight suits. Slater arranged for taking each of the pilots aloft on a parasail pulled by a Boston Whaler.

Colonel Slater, the flight surgeons Doctors Don Nichols and Don Vining, Castro, A-12 pilots Ken Collins, Jack Layton, Mele Vojvodich, Dennis Sullivan, and Frank Murray showed up at the lake wearing swimming trunks. Several others participated as well, especially the flight suit support personnel who suited up the pilots for the drop from the parasail.

The Coast Guard lifted fully suited pilots aloft in a parasail to a height sufficient for a parachute ride to the lake surface. The plan called for the pilot releasing the tow line, deploying his survival kit, and landing in the lake.

Col Hugh "Slip" Slater went first and did not get any water in the suit. Dennis Sullivan went second, and he too remained dry. After that, the boondoggle went to hell due to high winds preventing the little Boston Whalers from keeping up. Layton's helmet filled with water, causing him to think he was still under water. Murray slithered off to one side while Layton floundered around with his helmet full of water and Mele Vojvodich had a hard time deflating the chute. Slater canceled the exercise when the Whalers started having a hard time getting to the guys because the water had gotten rougher.

The parasail boondoggle did not develop any water survival techniques for a crew-member ditching in the ocean wearing a pressure suit. However, it did provide everyone a fun time and severe sunburns from frolicking in Lake Mead. Capt Charlie Trapp probably experienced the most fun while flying over the topless beaches with his helicopter calling "naughty, naughty" on the external speakers of the chopper.

CIA A-12 pilot Ken Collins training water survival in the Area 51 swimming pool with Capt Charlie Trapp and MSG Castro chatting on the side. *CIA vis TD Barnes Photo Collection*

One of the A-12 pilots waiting for the US Coast Guard Whaler to tow him over Lake Mead, Nevada for water survival training in his pressure suit. *Charlie Trapp*

Preparing A-12 pilot Jack Layton for water survival training in Lake Mead, Nevada. A US Coast Guard whale boat pulled the pilots airborne in a parasail where the pilot dropped into the water while wearing the pressurized suit normally worn on A-12

Capt Charlie Trapp was an excellent example of the little-known things occurring at Area 51 besides flight testing the A-12. He was sitting in the office of a helicopter rescue detachment in Charleston AFB, South Carolina when he received a call from rescue headquarters asking, "Do you want a job flying an H-43 helo at a classified location somewhere in the USA?" He was looking for some new thing to do so; he said yes. The caller said, "Find a helicopter and head west." Trapp asked their H-43 factory tech rep if he knew where to find an extra H-43. The tech rep told Trapp about one at Nellis AFB NV in the care of the local rescue unit. Trapped packed his family--wife and two kids--and headed out west.

He stopped by the helicopter rescue outfit at Nellis and looked at the H-43. The Nellis mechanics told Trapp that it was not in good flying condition. So, I flew it and confirmed it needed help.

Trapp then called the Area 51 Command Post to advise that he was in Las Vegas and had a problem. Col. Holbury got on the phone and asked what he needed and why. He informed Trapp that he would be supporting a high-altitude construction project and conducting rescue support operations. Trapp informed Holbury of the H-43 being in bad shape from snagging flight parachute payloads. He recommended that Holbury request a factory maintenance tech and a test pilot come to Nellis from the Kaman factory in Connecticut.

Trapp's first landing was on top of Mt. Baldy at 9,000 feet, the site of the construction project. The construction project on Mr. Baldy was a big one. It required the helicopter to haul in all personnel, equipment, food, water, parts, tower sections, wet cement, Conex box, welding equipment, and generators.

The contractor dug three holes which they filled with 30,000 pounds of wet cement transported by helicopter a half ton at a time. The cement contained four good-sized rods embedded to attach each pole base plate with matching holes.

Because of the heavy weight of each of the poles and the high-altitude, they had to reduce the fuel load with only the pilot on-board. They used a hundred feet of intercom cable so the crew chief could direct the pilot from the ground to raise the poles from the horizontal to the vertical position to place the poles on the concrete bases. Hovering was difficult with the high-density altitude, winds, and poor visual reference to the ground because of the steep sides of the mountaintop. Four ground people tethered the pole with ropes until they could secure the bolts to the base. It required running the helicopter engine at the redline for power during the lift.

Trapp quickly learned that everyone was in a hurry at Area 51. He trained firefighters as aircrew members on how to fight fires with the 1,000-pound fire bottles carried to the fire by helicopter dropping off the fire bottle and two firefighters close to the fire. The helicopter would then hover over the firefighters, cooling the area, and clearing the smoke away while they laid a foam trail to the simulated airframe to extract pilots and crew members from crash sites. Also, they checked them out on hoist operations and other crew duties.

Trapp trained the pararescue jumpers (PJs) as crewmembers and provided a platform to practice parachute jumps. Capt Keith Spencer, Crew Chief, MSG Walters, and he worked with contractors to plan the Mt. Baldy construction project. The helicopter was the only way to get equipment and people to the mountaintop construction site.

Setting up included establishing airborne alert procedures for the first few takeoffs and landings of the A-12. Normally, they provided a ground alert for A-12 ops and became airborne for declared emergency landings. When they first arrived at Area 51, they were asked to be airborne for A-12 takeoffs and landings. After a while, they decided that the A-12 was reliable enough so the helicopter could stand on ground alert.

Trapp quickly established a rescue and survival branch that included pilots, crew chiefs, firefighters, maintenance folks, and PJs. Their branch ran a survival school for the A-12 pilots and others, which included desert, jungle, seacoasts, and mountain training sites, parasail training at Lake Mead, water survival, and a pistol firing range.

Trapp's grouped worked with Lockheed, Para Jumpers, and Capt Charlie Cravatta's people on the contents of the A-12 seat survival kit.

Capt Cravatta, MSG Casto, and Trapp helped a Boulder company contractor put together a walk-around sleeping bag with overlapping pockets to retain heat. The completed bags were pressure packed to fit in the A-12 survival kit.

It was not long before Trapp began flying security checks around the perimeter of the Area and photo and security missions for the Atomic Energy Commission's underground testing events. The Indian Springs helicopters normally supported these tests, but when they were out of commission or transitioning to the new H-43s, they asked Trapp and his people at Area 51 to fill in. On one occasion, Trapp flew airborne alert for President Kennedy's Air Force One landing in Indian Springs. He also provided occasional rescue recovery operations for the Nellis Range aircraft accidents and bailouts.

Trapp's crews trained at Lake Mead doing water hoist pickups and PJ water jumps. On one occasion, MSG Casto jumped out of a C-130 at a high-altitude to test the A-12 pilots' pressure suit and was awarded a medal for his efforts.

Area 51 did get three more pilots, Ted Angle, Joe Pinaud, and Sam Scamardo. Spencer left after three years, became a Continental Airlines captain, and then retired. Angle left the Area sometime after June 1967 and became a test pilot at Edwards where he flew the U-2 and several other aircraft. Pinaud became a logistician and retired from Warner Robbins AFB (deceased 2009).

The extra pilots made more time to work on all the projects, flying VIPs and emergency missions in the U-3B, C-180, and C-210. Capt Trapp often flew Kelly Johnson, Jim Irwin (of Lunar astronaut fame), and many other VIPs to Burbank, Van Nuys, LAX, Edwards AFB, and Las Vegas.

Captain Trapp became the check pilot in the H-43, UH-1, U-3B, C-180, C-210, and co-pilot on the C-130. Trapp, like most of the CIA and air force personnel flying at Area 51 "fly all you want flying club," was soon flying co-pilot in the C-130 and even checking out in HC-130 rescue tanker.

The Area 51 Rescue and Survival Branch Members:

Helicopter Pilots	PJs	Crew Chiefs	Firefighters	
Charlie Trapp		Earl Casto	Bill (Flash) Walters	Unknown
Keith Spencer		Gordon Bailey	Tyndal	
Ted Angle		Fred Schneider	McCloud	

Joe Pinaud   Coy Staggs   Thomas Baker
Sam Scamardo

Trapp arrived as a captain in September 1962, promoted to major in April 1967, and departed Area 51 in June of 1967. He went on to become commander of seven air force units. Trapp was the 3rd Air Rescue Group Commander and on the staff of USSAG/7th AF staff in SEA 1975 at Nakhon Phanom Royal Thai Air Base, where they planned and participated in the evacuation of 287 non-combatants from the embassy in Phnom Penh, Cambodia, and the Ambassador and 6,000 non-combatants from Saigon. Two air force helicopter units, four of his staff, and he deployed out of Thailand to the aircraft carrier Midway where they helped plan and execute the evacuation of the non-combatants. Their helicopters also participated in the recovery of the United States civilian container ship Mayaquez from the Cambodians. He commanded two flying squadrons, the 41st ARRSq (HH-53s and HC-130s) and the 37th ARRSq (52 UH-1Fs, 105 pilots, and 275 enlisted); the 450-man 11th ARRS Consolidated Maintenance Squadron; and three HH-43 local base rescue units (two in SEA and one in the CONUS).

Trapp's aircrews were awarded ten air medals and several commendation medals for meritorious achievement while participating in aerial flights at the ranch. Charlie Trapp retired from the air force as a bird colonel.

Average A-12 flying time for the six Mach 3.2 qualified pilots increased to 353 hours. The training followed the same general pattern of the earlier groups of pilots for two new pilots entering the program in the fall of 1966. All A-12 pilots completed a Lockheed A-12 ground training course that consisted of normal and emergency procedures, propulsion and aircraft systems. They attended academic training concurrently with the A-12 transition flight program consisting of extensive instruction on the navigation, sensor, and ECM systems.

They devoted additional academic instruction to aircraft systems concurrently with operational readiness flight training and tactical doctrine study. The objective was to reach peak proficiency in all areas during the period. Standardization (ground phase) required the pilot to successfully pass a comprehensive examination on all aspects of aircraft operations and tactics. Integrated with the formal pilot training program were specialized courses of instruction in pressure suit fitting and physiological training, security and resistance to interrogation training, and escape and evasion training. Once declared operationally ready, each pilot assured optimum proficiency by receiving extensive ground and flying training.

To the maximum extent possible, OXCART project pilot training was consistent with flying safety and concurrent with the manufacturer's developmental flight testing. Lockheed incrementally cleared Detachment pilots to train at speed regimes that began at Mach 2.35 and advanced to Mach 3, the aircraft design operating speed in a unique training philosophy that proved advantageous. This way, the detachment attained operationally ready status far more quickly than had the normal air force Category I, II and III approaches,

All the OXCART pilot candidates went through A-12 flight training, already qualified in the F-101 aircraft to include the aerial refueling. They only used the A-12 trainer aircraft to augment pilot proficiency,

The basic A-12 flying training program consisted of 21 training missions. Eight of the flights were transition missions followed by a standardization check in the J75 equipped two seated A-12, trainer. The program required a total of twelve training flights in the J58-equipped A-12 during the operational readiness phase. Training included day and night transition flying, aerial refueling, instrument flying, photographic procedures, navigation, and ECM training. Their training included normal and emergency procedures for all systems, and during the final three training missions, they included simulated operational missions with multiple aerial refueling.

Opposition to Deployment and Use of OXCART.

The CIA attempts to commit the A-12's capability to operational use started even while the OXCART

Detachment was still in training for its functional role. At a 17 February 1966 meeting, the 303 Committee again questioned deploying the OXCART to the Far East. Secretary Cyrus Vance advised that his boss, Secretary of Defense Robert McNamara still opposed deployment on the basis that the situation had remained unchanged since December 1965.

On 22 March 1966, the director, Adm William F. Raborn expressed his serious concern over the lack of adequate photography. He felt the existing photography could not detect a possible Chinese Communist strategic build up in South China and China's direct involvement in the Vietnam war,

The inherent limitations of satellite coverage and the defensive threat to U-2s and drones severely inhibited the timely acquisition of high-resolution photography. He, therefore, recommended the immediate deployment of the OXCART aircraft to Okinawa and for the CIA to begin flying missions over North Vietnam as soon as possible, and be ready to back up the U-2 and satellite capability in South China should they prove incapable of fulfilling the intelligence requirements,

The joint chief of staff (JCS) supported this view. At the 28 March 1966 meeting of the 303 Committee, Secretary Vance reported that, despite the JCS views, and even though they felt the existing systems could provide sufficient coverage of North Vietnam, he and Mr. McNamara still opposed deployment. Since the Department of Defense was the key customer and was willing to live with lesser coverage, The Committee concluded not to deploy at that time. It was recommended, however, that after further study of aspects, bringing the views of the DOD and CIA to the attention of the president,

Subsequent discussions of the matter occurred on 11 May 1966 and 27 June 1966. The minutes of these meetings reflect that the CIA favored, immediate deployment and used the State Department was against the DOD was split, with Messrs. McNamara and /Vance opposed and the Joint Chiefs of Staff for implementation. The president's Foreign Intelligence Advisory Board (PFIAB) was on record as supporting implementation and use. The time had come to present the divergent views to the president. On 12 August 1966, Mr. Walt Rostow, after speaking with the president, advised the Director, Central Intelligence, Mr. Helms that the president had decided for the time being not to deploy the OXCART.

On 6 September 1966, a proposal was submitted to the 303 Committee to conduct OXCART reconnaissance missions over Cuba. The director had recommended, and the secretary of defense had concurred to exercising the OXCART capability over Cuba. The A-12 could confirm the reliability of the primary aircraft system and test the EWS against the Cuban SA 2 defenses if they responded. The stealthy A-12 would evaluate a Soviet-type defensive environment reaction and capability to a low cross-section, Mach 3.0 aircraft,

While it would provide higher resolution photography than the SAC's current and recently ordered U-2, The committee never proposed having the OXCART replace that capability. The primary objective was to establish and validate the operational capacities of the OXCART system to perform reconnaissance of a defended area,

On 15 September 1966, the attitude, in general, was negative when the 303 Committee considered the proposal. The opinion prevailed that despite their accruing advantages from their testing in a hostile environment, the CIA's introduction of the A-12 reconnaissance plane over Cuba might disturb the existing calm prevailing in that area of foreign affairs. The committee decided to wait until presented with an overriding requirement before the majority decided to commit the A-12 to a Cuban operation,

Further Development and Testing of A-12,

System Although the CIA had validated the A-12 performance at the high Mach and high-altitude, problems remained that it needed to work out. Until experiencing the real world of the A-12s flight regime, its effects on aircraft performance remained unknown. Vibrations, aeroelastic effects, thermal and mechanical shock during transients, systems interactions, control accuracies and environmental conditions made up this real world of high Mach flight,

One of the most troublesome problems was the inlet spike whose interaction affected range and posed a danger to the pilot and the aircraft. Its inaccuracy penalized inlet performance and caused what was known as unstarts. It increased fuel consumption by 5% to 10%, aircraft range by 5% to 10%,

Some problems affected the A-12's range. One was the engine turbine temperature that penalized the plane's range by 5 to 7% if the temperature reached 40°C low. Even the flight path environment of 10° C hotter than a standard flight penalized the aircraft's range by 7%. Temperature shears caused unscheduled control surface movements that resulted in additional drag. Even things such as the precise center of gravity management caused control surface movements resulting in additional drag. A deviation from optimum climb schedule reduced the fuel available for cruising. Overall, fuel temperature and density variations reduced fuel on load by 1,000 to 2,000 pounds.

The engineering development and flight testing concentrated on further improving the air inlet control system, main and afterburner controls, bypass door schedules, Mach hold, autopilot, and stability augmentation systems. OXCART management ran performance optimization flight program in conjunction with hardware improvements. By establishing a reliable cruise capability, that reflected an approximate 200-mile increase in the operational tanker to tanker range. The challenges and work paid off a flight carrying 6,000 pounds of fuel reserve could reach a cruising range of 2,870 nautical miles at altitudes ranging between 75,000 84,000 feet while traveling at Mach 3.1.

On 21 December 1966, a flight test aircraft would fly a two aerial refueling mission lasting 6:09 hours with three and a half hours flown at or above Mach 3.2. One leg of the mission covered 3,067 nautical miles, a range that the operational A-12s never attained because of program termination ground rules that restricted the necessary hardware improvements in the fleet. Though it remained doubtful that they could have achieved the original 4,000 nautical mile range figure, the technology had prospects of improving range performance beyond those demonstrated.

The Inverted Bailout

While on a routine flight, 24 May 1963, one of the detachment pilots, CIA pilot Ken Collins, Dutch 21, recognized an erroneous and confusing airspeed indication. He ejected from the aircraft, which crashed 14 miles south of Wendover, Utah. He escaped the event unhurt. The CIA, air force, and Lockheed recovered the wreckage in two days. The CIA identified all unauthorized observations of the crash site and obtained signed secrecy agreements. A cover story for the press described the accident as occurring to a Republic F-105 and remains listed in this way on official records. The CIA paid $25,000 cash to a rancher who came to Collin's aid to induce Area 51er to forget what he saw at the crash site.

The CIA grounded all A-12 aircraft for a week during the air force's investigation of the accident. A plugged pitot-static tube in icing conditions turned out responsible for the faulty cockpit instrument indications prompting Collins to jettison from the plane.

The loss of this aircraft nevertheless precipitated a policy problem which troubling the CIA. With the growing number of A-12s, how much longer could the project remain secret? The program had gone through development, construction, and a year of flight testing without attracting public attention. They experienced the Department of Defense having difficulty concealing its participation because of the increasing rate of expenditures, otherwise unexplained. There was also a realization of the technological data's extreme value as feasibility studies for the SST.

Finally, there existed a growing awareness in the higher reaches of the aircraft industry of something new and remarkable going on. Rumors spread, and gossip flew. Commercial airline crews sighted the A-12 in flight. The editor of Aviation Week (as might be expected) indicated his knowledge of developments at Burbank. The secrecy was thinning out.

In August 1963 Lockheed test pilot James "Jim" Eastham flew the first YF-12A (Article 1001/60–6934) on its maiden flight at Area 51. The YF-12 was one of three A-12s converted into a Mach 3 air force interceptor prototype armed with missiles.

The president's Announcement

Despite all this, 1963 went by without any public revelation. President Johnson was brought up to date on the project a week after taking office and directed the preparation of a paper for an announcement in the spring of 1964. Then at his press conference on 24 February, he read a statement announcing the United States has successfully developed an advanced experimental jet aircraft. He identified it as the A-11. A plane tested in sustained flight at more than 2,000 miles per hour and altitudes above 70,000 feet.

The president stated several A-11 aircraft are now flying at Edwards Air Force Base in California. Only three of these modified A-12 existed and had flown only at Area 51.

The president went on to mention the "mastery of the metallurgy and fabrication of titanium metal," giving credit to Lockheed and Pratt & Whitney. He remarked that appropriate members of the Senate and House were kept fully informed, and prescribed for keeping the full performance of the A-11 strictly classified.

The president deliberately referred to the "A-11," the original design designation for the all-metal aircraft first proposed by Lockheed. Subsequently, it became the design designation for the air force YF-12A interceptor, which differed from its parent mainly in that it carried a second man for launching air-to-air missiles.

To preserve the distinction between the A-11 and the A-12, Security had briefed all waiting personnel in government and industry on the impending announcement practically. OXCART secrecy continued in effect. There was considerable speculation about an Agency role in the A-11 development. However, the government never acknowledged it. News headlines ranged from "US has dozen A-11 jets already flying" to "Secret of sizzling new plane probably history's best kept."

On February 29, 1964, five years after work began on the A-11, President Lyndon Johnson finally held his long-desired press conference. He told reporters of the aircraft attaining speeds of over 2,000 mph and altitudes of more than 70,000 feet during tests at Edwards Air Force Base.

By now, the A-11 and advanced to the A-12 production version with a reduced radar cross-section.

In October 1960, the air force contracted for three interceptor versions of the A-12, which now became available. President Lyndon Johnson got word of the Mach 3 planes and decided to announce this historic aeronautical benchmark to enhance his election for a second term. He approached the CIA first, wanting them to exhibit the top-secret A-12. The CIA flatly refused to do so, telling the president that if he wanted to expose the US having such a plane, he could reveal the planes flown by the air force.

No Blackbirds had ever flown at Edwards AFB when he spoke, announcing the air force flying the planes at Edwards. Project officials knew about the impending public announcement. However, none knew exactly when. Caught by surprise, they hastily flew two air force YF-12A's to Edwards to support the president's statement. Rushed by the surprise announcement, the air force rushed the planes directly into hangars upon arrival. The heat from the jet engines and the fuselage activated the hangar sprinkler systems, dousing the reception team that awaited them.

Thenceforth, while the A-12 continued its secret career at its site, the A-11 performed at Edwards Air Force Base in a much glare of publicity. Pictures of the aircraft appeared in the press; correspondents could look at it and marvel at stories written, yet they learned virtually no details. Nonetheless, the technical journals nevertheless enjoyed a field day. The unclassified Air Force and Space Digest, for example, published a lengthy article in its issue of April 1964, stating: "The official pictures and statements tell very little about the A-11. However, the technical literature from open sources, when carefully interpreted, says a good deal about what it could and, more importantly, what it could not be. Here's the story…"

The president's exposing the YF-12 left Kelly Johnson no choice but move the YF-12A test program to Edwards AFB, California.

Intruders Are Not Welcome

In October 1963, a flight of three F-105 Thunderchiefs, led by British exchange pilot Anthony "Bugs" Bendell was flying a practice nuclear weapon delivery sortie. The flight was about 80 miles north of Nellis AFB when one of the planes experienced an oil pressure malfunction. One F-105 returned to Nellis while Bendell led the stricken craft to what the air force pilots referred to as the box. The no-fly zone around Groom Lake showed as a square on their maps.

The planes made a pass over the field with no response to distress calls. Bendell advised the student pilot to land.

At this point, two F-101 Voodoos intercepted Bendell and forced him to land. The CIA and Air Force at Area 51 held the pilots there for two weeks of debriefing to discourage other pilots from landing at Groom Lake.

The CIA finally released him to Nellis AFB authorities, however, refused to release the plane for several more weeks. The air force got the CIA's message and stopped all future emergency landings by young pilots wanting to see what was in the box.

Going Operational

Three years and seven months after first flight in April 1962, Lockheed and the CIA declared the A-12 ready for operational use at design specifications. The period thus devoted to flight tests was remarkably short, considering the new fields of aircraft performance under exploration. As the A-12s reached each higher Mach number, the support contractors continued correcting defects and making improvements. Everyone concerned gained experience with the characteristics and idiosyncrasies of the vehicle.

With approximately 1000 individuals assigned, a mix of air force, CIA, many varied contractors such as Car Co, David Clark, EG&G, EK Kodak, Firewel, Honeywell, Magnavox, Perkins Elmer, Pratt & Whitney (aka United Aircraft Co), Ree Co, Sylvania (Blue Dog ECM C&J Engineering (parts supply, Hughes, Hamilton Standard, and of course Lockheed personnel all being involved in some manner or form. That these companies were involved, or what services they performed was probably never known by the clear majority of those assigned to the Area.

The air inlet and related control continued for a long time to present the most troublesome and refractory problem. Numerous attempts failed to find a remedy, even though a special task force concentrated on the task. For a time, there was something approaching despair, and the solution, when finally achieved, they greeted with enormous relief. After all, not every experimental aircraft of advanced performance has survived its flight testing period. The possibility existed of OXCART also failing despite the significant cost and effort expended upon it.

A few dates and figures will serve to mark the progress of events. The year 1963 ended with 573 flights totaling 765 hours, and nine aircraft in the inventory.

On 20 July 1963, the A-12 flew for the first time at Mach 3. In November, it reached Mach 3.2, the design speed, and reached 78,000 feet altitude. The longest sustained flight at design conditions occurred on 3 February 1964; lasting ten minutes at Mach 3.2 and 83,000 feet. The end of 1964 totaled 1,160 flights and 1,616 hours flight time with eleven aircraft available, four of them reserved for testing and seven assigned to the detachment.

Stating the record another way, the A-12 reached Mach 2 after six months of flying; Mach 3 after 15 months. Two years after the first flight the aircraft flew a total of 38 hours at Mach 2, three hours at Mach 2.6, and less than one hour at Mach 3. After three years, Mach 2 time increased to 60 hours, Mach 2.6 to 33 hours, and Mach 3 time to nine hours. Only the test aircraft flew any Mach 3 time. The detachment aircraft remained restricted to Mach 2.9.

As may be seen from the figures, most flights lasted a short duration, averaging little more than an hour each, longer flights unnecessary at this stage of testing. Everyone felt the less seen of the A-12, the better, and short flights helped to preserve the secrecy of the proceedings. It remained virtually impossible for an aircraft of such dimensions and capabilities to remain inconspicuous. At its full speed, OXCART required a turning radius of no less than 86 miles, and at times up to 125 miles. There was no question of staying close to the airfield; its shortest possible flights took it over a vast expanse of territory.

The first long-range, high-speed flight occurred on 27 January 1965. One of the test aircraft flew an hour and fifteen minutes above Mach 3.1 for 2,580 nautical miles total range, at altitudes between 75,600 and 80,000 feet.

The year 1965 saw the test site reach the high point of activity with all the detachment pilots Mach 3.0 qualified. Completion of construction brought it to full physical size with a site population of 1,835. Contractors worked three shifts a day. Lockheed Constellations flew daily flights between the factory in Burbank and the site. The C-47 flew two flights a day between the site and Las Vegas. Now officials began considering how and when and where to use OXCART in its appointed role.

A-12s Lost at Area 51

Following Collins' crash, the program during this phase lost two more aircraft. On 9 July 1964, Article No. 133 made its final approach to the runway. At an altitude of 500 feet and airspeed of 200 knots, it began a smooth, steady roll to the left. Lockheed test pilot Bill (Dutch 50) could not overcome the roll. At about a 45-degree bank angle and 200-foot altitude he ejected. He swung down to the vertical in the parachute at the same time his feet touched the ground, for what must have been one of the narrower escapes in the perilous history of test piloting.

The primary cause of the accident was a frozen servo for the right outboard roll and pitch control. No news of the crash ever filtered out.

On 28 December 1965, Aircraft No. 126 crashed immediately after takeoff and was destroyed. Detachment pilot Mele Vojvodich (Dutch 30) ejected safely at an altitude of 150 feet. Like what happened to Bill Park, Vojvodich's parachute opened at the same time his feet touched the runway. He suffered a sprained ankle.

The accident investigation-board determined that a flight line electrician had improperly connected the yaw and pitch gyros had in effect reversed the controls. This time Mr. McCone directed the Office of Security to investigate the possibility of sabotage. While discovered no evidence of sabotage, they found indications of negligence. The manufacturer of the gyro earlier warned Lockheed of the possibility of connecting the mechanism could engage in reverse. No one acted or even an elementary precaution such as painting the contacts different colors. Again, no publicity occurred related to the accident.

Besides the pilot narrowly escaping death, the accident proved spectacular in another way. The A-12 aircraft required a special fuel with a high flash point and thermal stability. The fuel, JP-7 (Jet Propellant 7), required a radioactive Cesium additive as a stealth feature to reduce the radar signature of its exhaust plumes. Everyone referred to the additive as Panther Piss.

The fuel also used triethylborane (TEB) to ignite the engines. This additive ignited when it met the air. TEB produced a characteristic green flame seen during engine ignition. When Mele Vojvodich's plane crashed on the runway at Groom Lake, ice covered the runway. The crash released the triethylborane that ignited as it spread beneath the ice. By the time the rescue vehicles could respond, the burning TEB covered a large area beneath the ice near the crash site.

At the time of Vojvodich's crash, Col Slip Slater (Dutch 11), commander of the 1129th SAS at Groom Lake was in California visiting his daughter during the Christmas holiday, leaving Colonel Holbury in command at the facility. Maj Harold Burgeson was on duty at the Ops building when the accident occurred. Hearing that Mele had just crashed, he headed for the Ops vehicle at a dead run. Just as he reached the outside gate, Col Holbury screeched to a halt in his staff car, picked him up, and they went to the site together. After assuring that Vojvodich was OK, they looked at the wreckage before going to see him. The aircraft was grossly out of trim. Project Test Pilot, Denny Sullivan was in another station wagon monitoring the take-off and narrowly missed Mel when he drove to the crash site.

Maj Roger Andersen was on duty in the command post monitoring the tower frequencies during take-off when he heard the aircraft crash and rushed out on the lakebed. He saw Vojvodich land quite close to where the plane crashed. He witnessed one of the fire trucks narrowly miss running over Vojvodich in its rush to get to the fire.

The accident occurred about dusk, and the fire trucks arrived on the scene quickly. One of the fire trucks rushed quite close to Vojvodich standing and watching the thick black smoke and orange flames boiling from the wreckage. The fire trucks gained control of the flames coming from the wreckage. About that time, Andersen saw fuel from the crash area flowing out onto the lakebed and getting under a thin layer of ice. The TEC ignited on its own and continued burning under the ice with an eerie greenish white flame, looking like a large votive candle as darkness set in at Groom Lake.

Vojvodich merely sprained his ankle when he bailed out, escaping death by inches. When he returned home to Los Angeles, his wife, Carol, asked him about his limping. He told her that he sprained his ankle playing tennis.

Major Burgeson served as a member of the accident board where the Lockheed team determined the SAS connections accidentally reversed, causing the plane to misinterpret the pitch and yaw signals. A few days later, base commander Colonel. Slater, the project pilots, Major Burgeson, and Lockheed test pilot Bill Park went to Beale AFB to check the cable reversal out in Beale's new simulator for the SR-71. Mele Vojvodich and a colonel from Wright-Patterson AFB accompanied them.

Bill Park took the first flight in the simulator with the cables reversed while the rest waited in an adjacent room. Bill Park had a tremendous sense of humor, and when he returned, he winked at Burgeson then remarked that it was a rough ride, however, flyable. Burgeson then took his flight, and when he returned, he continued the charade with a similar observation. Per Burgeson, Vojvodich looked so crestfallen, that they burst into laughter and confessed to both of them crashing in the simulator.

What the Historians Never Knew About Vojvodich's Crash:

During Project OXCART, one of the air force and CIA's most loyal defenders of their careers was BGen Jack C. Ledford. On August 1958, General Ledford received an assignment to deputy chief of staff for weapons effects and tests, Headquarters, Defense Atomic Support Agency, Washington, DC.

He left this position in 1961 to attend the Industrial College of the Armed Forces, graduating with distinction in August 1962. Washington during his tour of duty, he earned a master of business administration in management from The George Washington University.

In September 1962, he served as an air commander with the 1040th Air Force Field Activity Squadron at Bolling Air Force Base, Washington, DC. Following that, he became the Director of special projects, Headquarters Air Force, making him the director of the Office of Special Activities, DD/S&T for Project Oxcart at Area 51.

Ledford was newly assigned to the CIA and leading OSA in 1962 when he took the request to the Special Group to get authorization for the 14 October 1962 flight over Cuba. He met stiff resistance, however, held his ground against the "do-nothing, worry a lot crowd." Everyone acted apprehensive after a SAC U-2 strayed slightly over Russia and the CIA lost a U-2 over China that summer.

Bobby Kennedy came to his rescue and insisted on a vote up or down. The Special Group approved the mission and caught the Russian missiles in Cuba. The rest was history.

When Mele Vojvodich bailed out on takeoff, Jack Ledford, and Dr. Albert D. "Bud" Wheelon, Ph.D., Director for Science and Technology at CIA immediately flew to Los Angeles and picked up Kelly Johnson en route to Area 51.

On the way up, Johnson started bitching about the quality of the CIA's operational pilots. General Ledford took issue, and it ended up with Wheelon breaking up a fistfight between General Ledford and Kelly Johnson in the plane's small cabin as they headed to Groom Lake.

Ledford always stood up for his people and good reason. It turned out that Kelly Johnson's people caused the crash by inserting the two augmentation rate gyros in backward on Vojvodich plane. (The author, TD Barnes (Thunder), CIA pilot Frank Murray (Dutch 20), and Roger Andersen attended General Ledford's funeral in Tucson, Arizona. Dr. Wheelon attended as well and told this story as part of his eulogy for the general.

While about fuels, another occurrence comes to mind. In 1958, Shell Oil vice president Jimmy Doolittle arranged for the company developing the fuel for the Central Intelligence Agency's secret Lockheed A-12 spy plane. The A-12 needed a low-volatility fuel that wouldn't evaporate at high-altitude. Manufacturing several hundred thousand gallons of the new fuel required the petroleum byproducts Shell commonly used to make Flit insect repellant, causing a nationwide shortage of that product that year.

In July 1966, Lockheed made a fourth launch attempt from M-21 (60–6941) with 60–6940 flying chase. Thus, the second A-12 converted to an M-21 for launching the D-21 departed Groom Lake on 30 July 1966. Lockheed Test Pilot Bill Park piloted the aircraft, and Lockheed engineer Ray Torrick the LCO flew back seat for launching the drone.

The mothership and drone went feet wet at Point Mugu, California and launched the drone over the Pacific. Following the launch, the drone pitched down and struck the M-21 Mothership, breaking it in half. Pilot Bill Park and LCO (Launch Control Officer) Ray Torrick stayed with the plane a short time before ejecting over the Pacific Ocean. Both made a safe ejection. However, Ray Torrick opened his helmet visor by mistake, and his suit filled up with water which caused him to drown.

This terrible personal and professional loss drove Kelly Johnson to cancel the M-21/D-21 program. This accident also prompted water survival training by the A-12 pilots based at Groom Lake. Under the supervision of 1129th SAS commander, Col Hugh Slater, the pilots, wearing their flight suits, lifted high above the waters of Lake Mead on a parasail towed by a United States Coast Guard whaler. Colonel Slater quickly aborted the training when some of the fully suited pilots almost drowned after dropping from the parasail into the water.

In January 1967, while returning to Area 51 from a routine training flight, A-12 Article #125 crashed near Leith, Nevada. A faulty gauge had allowed the jet to run out of fuel 70 miles short of Groom Lake. Walt Ray (Dutch 45) ejected, however, failed to separate from his seat, killing him on impact with the ground. Walt Ray had married only three months earlier.

Capt Charlie Trapp was in Las Vegas with the UH-1 helicopter when he received a call that Walt Ray was missing. He gathered his crew and a Nellis flight surgeon and started the search. Dark mountain terrain and high winds made it difficult. The H-43 searched from Area 51. They called off the search because the H-43 was low on fuel and there was no sign of fire nor were emergency radio signals heard. They rescheduled first light ops for the next day. He took an F-105 pilot and flight surgeon with him and his Pararescue jump crew. The F-105 pilot thought he knew the approximate crash location.

They flew the UH-1 to the area and found the A-12 very soon after arriving. The drag chute had deployed. They deployed the PJs to search the area and the aircraft and discovered that the ejection seat and Walt were missing.

They searched the rest of the day without success and planned the next day's search with several agencies. They started at the point of impact and searched back along the flight path. Trapp's UH-1 crew flew past Ray that morning and did not see him due to shadows caused by the low sun angle. Later in the day, the C-47 crew saw a sun reflection from his suit or visor, and they directed us to the site where they landed and picked up Ray. He was still in his ejection seat. They took him to Nellis and his new bride, Diane.

They later returned to search for the canopy and the camera--it took 15 days to find them using horses as well.

A month or two before all this happened, the PJs and Captain Trapp had taken Walt Ray and other pilots to Fort Myers Florida for jungle and sea coast survival training. The only transport to the training site was by boat. The guys had to survive on land vegetation, fish, and turtles for several days. Trapp provided the psychological stress by announcing at the end of each day that he was leaving for the night for a shower, a few drinks, and a steak dinner and they were not. They chased him to the boat, but all was forgotten when he brought them a case of cold beer for the last night.

The Loss of Pilots Continued

On 27 September 1967, James S. Simon, Jr., flew chase for a night sortie of the TA-12, the two-seat A-12 trainer affectionately called the Titanium Goose, flown by CIA pilot Jack Layton and air force pilot Harold Burgeson. As the TA-12 approached the south end of the runway, Simon's F-101B struck the ground and exploded near the south rim pad.

Layton, Dutch 12, the pilot occupied the front cockpit and Burgeson, the instructor pilot the rear. Jim Simon, the chase pilot, flew the F-101B (56–0286). After the trainer had become safely airborne and all systems checked normal, Simon routinely flew around in the local area to await their return.

Layton and Burgeson completed their mission and returned to Groom Lake where Layton started an instrument letdown for a full stop landing. During the letdown, Burgeson called the chase and informed Simon of their return with no problems. Simon asked for their position, and Burgeson gave it to him, again stating all systems normal and the A-12 trainer not needing any assistance. Simon responded that he wanted to find them at least.

Layton turned the trainer on final approach and received clearance for a full stop landing. In the cockpit, neither could physically see the wings or engines. On short final, a sudden explosion occurred off the trainer's right wing. Layton and Burgeson saw the flash and felt the concussion. Layton instinctively stop cocked the right engine, lit the left afterburner, and said, "Burgie be ready to bail out." Burgeson replied, "That was not us, Jack. It was the chase. If you keep this thing flying straight, I will restart the right engine." Burgeson got the engine started, and they circled for a landing. Both avoided looking at the fire as they approached the runway and Layton made a smooth landing. Until they called for landing clearance, the tower operators thought it was A-12 trainer aircraft that crashed.

No one ever knew for sure what caused the crash. What the pilots knew was that joining up with a dark, unlit airplane on a night at final approach airspeed is not easy. The aircraft contacted the ground in a flat attitude near the South Trim Pad of the Groom Lake landing strip. The manner of crashing suggests that Simon got a little low and flew into the ground. They could only speculate that he might have overshot a little and dropped down for clearance or that something in the cockpit, such as a warning light distracted him. This was speculation that served no useful purpose. In any case, that night the OXCART project lost an exceptional officer, an excellent pilot, and a good friend. Simon left a spouse and three sons. Simon's widow never remarried and remained in their Las Vegas home until she died in 2006.

Also in 1967, Operation BLACK SHIELD meteorologist Weldon "Walt" King was TDY from Groom Lake to Kadena when killed in an F-101 VooDoo during a weather flight. He met with bad weather during which he lost the tail of his aircraft. King survived bailing out only to have the plane crash on top of him.

The air force lost the third YF-12A on 24 June 1971 in an accident at Edwards AFB when a fire broke out while Lt Col Ronald J. Layton and systems operator William A. Curtis approached the traffic pattern. A fuel line fracture caused by metal fatigue enveloped the entire aircraft in flames on the base leg, forcing both crew members to eject from Article 936 only moments before it crashed and burned near Barstow, California. The article still bore a large, white cross painted on its belly from when it made the speed runs in 1965.

An impressive demonstration of the A-12's capability occurred on 21 December 1966 when Lockheed test pilot Bill Park flew 10,198 statute miles in six hours. The aircraft left the test area in Nevada and flew northward over Yellowstone National Park, thence eastward to Bismarck, North Dakota, and on to Duluth, Minnesota, where it then turned south and passed Atlanta en route to Tampa, Florida. The plane turned northwest and headed towards Portland, Oregon and then southwest to Nevada. Again, the flight turned eastward, passing Denver and St. Louis. Turning around at Knoxville, Tennessee, it passed Memphis in the home stretch back to Nevada. This flight established a record unapproachable by any other aircraft; it began at about the same time a typical government employee started his work day and ended two hours before his quitting time. *

Tragedy befell the program during a routine training flight on 5 January 1967 when the fourth aircraft was lost, together with its pilot. The accident occurred during descent about 70 miles from the base. A fuel gauge failed to function properly, and the aircraft ran out of fuel only minutes before landing.

The pilot, Walter Ray, ejected but died when he failed to separate from the ejection seat before impact. The air force located the wreckage on 6 January and recovered Ray's body a day later.

Through air force channels, the air force released a story to the effect that an air force SR-71, on a routine test flight out of Edwards Air Force Base, was missing and presumed down in Nevada. The announcement identified the pilot as a civilian test pilot, and the newspapers connected him with Lockheed. Flight activity at the base again suspended during an investigation of the causes both for the crash and for the failure of the seat separation device.

It is worth observing that none of the four accidents occurred in the high Mach number, the high-temperature regime of flight. All traditionally involved problems inherent in any aircraft the OXCART was by this time performing at high-speeds, with excellent reliability.

Chapter 10 -- Operation BLACK SHIELD

In 1965, a more critical situation did indeed emerge in Asia and interest in using the aircraft there manifested. On 18 March 1965, Mr. McCone discussed with Secretaries McNamara and Vance the increasing hazards to U-2 and drone reconnaissance of Communist China. A memorandum of this conversation stated:

"It is further agreed they proceed immediately with all preparatory steps necessary to operate the A-12 over Communist China, flying out of Okinawa. They agreed the CIA should proceed with all construction and related arrangements. However, this decision did not authorize the deployment of the A-12 to Okinawa nor the decision to fly the A-12 over Communist China. The decision allows all preparatory steps and the expenditure of such funds as might be involved. No one decided to fly the A-12 operationally over Communist China. Only the president can make this decision."

Four days later, BGen Jack C. Ledford, director of the Office of Special Activities, DDS&T, briefed Mr. Vance on the scheme drawn up for operations in the Far East. The project code-named BLACK SHIELD called for the A-12 operating out of the Kadena Air Force Base in Okinawa. The first phase staged three aircraft Okinawa for 60-day periods, twice a year, with about 225 personnel involved.

Once established, BLACK SHIELD was supposed to become a permanent detachment at Kadena. Secretary Vance made $3.7 million available for providing support facilities on the island by early fall of 1965. A year later, BLACK SHIELD ended as the air force replaced the CIA flying surveillance in SEA and all the other war zones for the next two decades.

Meanwhile, the Communists deployed surface-to-air missiles around Hanoi, thereby threatening the current military reconnaissance capabilities of the United States. Secretary McNamara called this to the attention of the undersecretary of the Air Force on 3 June 1965 and inquired about the practicability of substituting the A-12s for the U-2s. The CIA told him that BLACK SHIELD could operate over Vietnam once it achieved adequate aircraft performance.

With deployment overseas, thus apparently impending in the fall, the detachment went into the final stages of its program for validating the reliability of aircraft and aircraft systems. The A-12 set out to demonstrate total system reliability at Mach 3.05 and 2,300 nautical mile range, with a penetration altitude of 76,000 feet. The validation process included a demonstrated capability for three aerial refueling.

By this time, the A-12 was well along in performance. The inlet, camera, hydraulic, navigation, and flight control systems all demonstrated acceptable reliability. Nevertheless, longer flights conducted at high speeds and high temperatures surfaced new problems, mostly with the electrical wiring system.

Wiring connectors and components withstood temperatures of more than 800 degrees Fahrenheit, together with structural flexing, vibration, and shock. Wiring failures and careless handling attributed to continuing malfunctions in the inlet controls, communications equipment, ECM systems, and cockpit instruments many cases. Difficulties persisted in the sealing of fuel tanks. Because of these problems, officials feared meeting the scheduled date for BLACK SHIELD readiness. Prompt corrective action by Lockheed was in order. The quality of maintenance needed drastic improvement. The responsibility for delivering an aircraft system with acceptable reliability to meet an operational commitment lay in Lockheed's hands.

In this uncomfortable situation, John Parangosky, deputy for Technology, OSA, went to the Lockheed plant to see Kelly Johnson on 3 August 1965. A frank discussion ensued on the measures necessary to ensure that BLACK SHIELD met its commitments, and Johnson concluded that he should himself spend full time at the site to get the job done expeditiously. Lockheed President Daniel Haughton offered the full support of the Corporation, and Johnson began duty at the site the next day. His firm and effective management got project BLACK SHIELD back on schedule.

The four primary BLACK SHIELD aircraft selected conducted their final validation flights. During these tests, the A-12 achieved a maximum speed of Mach 3.29, the altitude of 90,000 feet, and sustained flight time above Mach 3.2 of one hour and fourteen minutes. The maximum endurance flight lasted six hours and twenty minutes. They reached the final stage on 20 November 1965, and two days later, Kelly Johnson wrote General Ledford:

"Overall, my opinion is that we can successfully deploy the aircraft for the BLACK SHIELD mission with what I consider at least as low a degree of risk as in the early U-2 deployment days. Considering our performance level of more than four times the U-2 speeds and three miles more operating altitude, it is probably much less risky than our first U-2 deployment. I think the time has come when the bird should leave its nest."

Ten days later the 303 Committee received a formal proposal that deployed OXCART to the Far East. The committee examined the matter but did not approve it, agreeing, however, that short of moving aircraft to Kadena, the CIA should take all steps to develop and maintain a quick reaction capability. The committee wanted the A-12 ready to deploy within a 21-day period at any time after 1 January 1966.

There the matter remained for more than a year. During 1966, the 303 Committee received repeated requests for authorization to deploy OXCART to Okinawa and conduct reconnaissance missions over North Vietnam, Communist China, or both. The committee turned them all down.

The First Demise of Project OXCART

"Give the OXCART mission to SAC and dispose of the A-12 aircraft" Gen Curtis LeMay had told the CIA even before it built the U-2 that once the CIA built and paid for the U-2, his Strategic Air Command would take it away. He was working at that, and now he wanted to do the same with the CIA's A-12 with the politicians backing him. Among high officials, there was a difference of opinion. The CIA, the Joint Chiefs of Staff, and the president's Foreign Intelligence Advisory Board favored the move, while Alexis Johnson, representing State, and Defense in the persons of Messrs. McNamara and Vance, opposed it.

The proponents urged the necessity of better intelligence, on a possible Chinese Communist buildup before intervention in Vietnam. The opponents felt no urgent need existed for better intelligence to justify the political risks of basing the aircraft in Okinawa. They feared this disclosing the A-12's existence to Japanese and other propagandists. They also believed it undesirable to use OXCART and reveal something of its capability until a more urgent requirement appeared. At least once, on 12 August 1966, they presented the divergent views to the president, who confirmed the 303 Committee's majority opinion against deployment.

The rumors of the cancellation of Project OXCART and the air force's SR-71 taking over had a devastating effect on Area 51. Many of the air force and civilian support as well began seeking a place to land should Washington pull the plug on Area 51. Many of the EG&G special projects engineers and technicians sought their former positions on the Nevada Proving Grounds, costing the CIA its technical capabilities of Area 51. It also cost the CIA one of its A-12 pilots, Bill Skliar who asked to return to the air force to fly the YF-12 at Edwards AFB, California.

A study group including C. W. Fischer, Bureau of the Budget; Herbert Bennington, Department of Defense; and John Parangosky, Central Intelligence Agency conducted a study through the fall of 1966 and identified three principal alternatives of its own, i.e.,

1. To maintain the status quo and continue both fleets at current approval levels.

2. To mothball all A-12 aircraft and maintain the OXCART capability by sharing SR-71 aircraft between SAC and CIA.

3. To terminate the OXCART A-12 fleet in January 1968 (assuming an operational readiness date of September 1967 for the SR-71) and assign all missions to the SR-71 fleet.

John Parangosky

On 12 December 1966, there was a meeting at the Bureau of the Budget attended by Mr. Helms, Mr. Schultze, Mr. Vance, and Dr. Hornig, Scientific Advisor to the President. Those present voted on the alternatives proposed in the Fischer-Bennington-Parangosky report. Messrs. Vance, Schultze, and Hornig chose to terminate the OXCART A-12 fleet, and Mr. Helms stood out for the future sharing of the SR-71 fleet between CIA and SAC.

The Bureau of the Budget immediately prepared a letter to the president setting forth the course of action recommended by the majority. Mr. Helms, having dissented from the majority, requested his deputy director for Science and Technology to prepare a letter to the president stating CIA's reasons for remaining in the reconnaissance business.

On 16 December, Mr. Schultze handed Mr. Helms a draft memorandum to the President who requested a decision either to share the SR-71 fleet between CIA and SAC or to terminate the CIA capability entirely. This time Mr. Helms replied that he had received any new information of significance concerning SR-71 performance.

He requested another meeting after 1 January to review pertinent facts and asked for withholding the memorandum to the president pending that meeting's outcome. Specifically, he cited indications of the SR-71 program experiencing severe technical problems and real doubt of achieving an operational capability by the time suggested for termination of the A-12 program.

Mr. Helms, therefore, changed his position from sharing the SR-71 aircraft with SAC to a firm recommendation to retain the OXCART A-12 fleet under civilian sponsorship. The president received transmittal of the Budget Bureau's memorandum. On 28 December 1966, he accepted the recommendations of Messrs. Vance, Hornig, and Schultze, and directed the termination of the OXCART Program by 1 January 1968.

This decision meant developing a schedule for orderly phase-out. After consultation with project headquarters, the Budget Bureau advised the deputy secretary of defense of the termination program on 10 January 1967. The termination scheduling placed four A-12's in storage in July 1967, two more by December, and the last four by the end of January 1968. Colonel Slater and his deputy, Colonel Anderson rebuilt the 1129th much easier than did the CIA its technical people lost from Special Project. This time, the CIA told EG&G that it did not want the company to man its special projects team with people borrowed from the atomic testing next door. The CIA resupplied the special projects technical and professional ranks by picking whom it wanted on the special projects team and told EG&G to carry them on its special activities payroll.

The extension of the termination date for OXCART meant the CIA and the Air Force's 1129th SAS must bring back the personnel lost to new job jobs or to hire replacements.

In May, Mr. Vance directed that the SR-71 assume contingency responsibility to conduct Cuban overflights as of 1 July 1967, and take over the dual capability over Southeast Asia and Cuba by 1 December 1967, providing for some overlap between OXCART withdrawal and SR-71 assumption of responsibility.

Meanwhile, until 1 July 1967, the OXCART Detachment maintained its capability to conduct operational missions both from a prepared location overseas and from the US. The plan included a 15-day quick reaction capability for deployment to the Far East and a seven-day immediate response for implementation over Cuba. Between 1 July and 31, December 1967 the fleet was to remain able to conduct operational missions either from a prepared overseas base or home base, however, not from both simultaneously. A quick reaction Project OXCART suffered through two demises. The first occurred during November 1965, the very month that the CIA and Lockheed finally declared the OXCART A-12 operational; Washington made the beginning move toward its demise. Within the Bureau of the Budget, a circulating memorandum expressed concern at the costs of the A-12 and SR-71 programs, both past and projected.

Gen Curtis LeMay

It questioned the requirement for the total number of aircraft represented in the combined fleets and doubted the necessity for a separate CIA OXCART fleet. Several proposed alternatives achieved a substantial reduction in the forecasted spending. The recommended course of action phased out the A-12 program by September 1966 and stopped any further procurement of SR-71 aircraft.

Copies of this memorandum went to the Department of Defense and the CIA with the suggestion that those agencies explore the alternatives set out in the paper. However, the secretary of defense declined to consider the proposal, presumably because the SR-71 is not becoming operational by September 1966.

Things remained in this state until July 1966 when the Bureau of the Budget proposed establishing a study group to consider the possibility of reducing expenses on the A-12 and SR-71 programs. The Bureau of Budget requested the group to consider the following alternatives:

1. Retention of separate A-12 and SR-71 fleets, i.e., status quo.
2. Collocation of the two fleets.
3. Transfer of the OXCART mission and aircraft to SAC.
4. Transfer of the OXCART mission to SAC and storage of A-12 aircraft
5. Transfer of the capability for either Cuban overflights or deployment to the Far East also is maintained.

All these transactions and arrangements had occurred before the A-12 had conducted a single operational mission or even deployed to Kadena for such a mission. As recounted above, the aircraft first performed its appointed role over North Vietnam on the last day of May 1967.

In succeeding months, it demonstrated both its exceptional technical capabilities and the competence of its operations management. Word circulated about phasing out OXCART, causing the program's high officials to feel some disquiet. These included Walt Rostow, the president's special Assistant; by key Congressional figures, members of the president's Foreign Intelligence Advisory Board, and the president's Scientific Advisory Committee. The phase-out lagged, and the question reopened concerning OXCART's future.

The committee completed a new study of the feasibility and cost of continuing the OXCART program in the spring of 1968 and proposed four new alternatives.

1. Transfer all OXCART aircraft to SAC by 31 October 1968; substitute air force for contractor support where possible; turn the test A-12 aircraft over to the SR-71 test facility.
2. Transfer OXCART as in Alternative 1, above, and store eight SR-71s.
3. Close the OXCART home base and collocate the fleet with SR-71s at Beale Air Force Base in California with CIA retaining control and management.
4. Continue OXCART operations at its base under CIA control and management.

The Loss of Article 125 and CIA Pilot Walter Ray.

On 5 January 1967, the fourth A-12 was lost. Article 125's accident occurred during descent about 70-miles from Area 51. While flying near Leith, Nevada, the pilot, Walter "Walt" Ray experienced a fuel system gaging malfunction which resulted in a higher than actual fuel quantity reading. Because of this, the aircraft's fuel supply depleted before reaching the base, destroying the plane. Ray, married only three months, ejected but died when he failed to separate from the ejection seat before its impact. Area 51 personnel and aircraft initiated a search for the wreckage and the pilot.

The search party discovered the crash site on 6 January and recovered Ray's body on 7 January. A cleanup crew spent a week retrieving pieces of the wreckage.

The cover story for the accident was that an SR 71 aircraft out of Edwards A FB was missing and presumed down in Nevada.

The aircraft was on a routine test flight, and the pilot was missing. The air force released this story via air force channels on 6 January and identified the pilot the next day as a civilian test pilot. Although the air force statement did not disclose Ray's employer, the newspapers identified him as a Lockheed employee, a story never refuted by Air Force Public Information Officers.

Upon determination of causes, precautionary action in the form of inspections and calibration checks was taken to preclude recurrence of similar malfunctions. It is interesting to note the fact that none of the four A-12 accidents involved the high Mach number high, temperature, the regime of flight. All the accidents involved traditional problems inherent in any aircraft. In fact, the A-12 by now was performing with a high degree of reliability at the high Mach environment. Detachment flight sortie effectiveness was near 80%. Major subsystem reliability was also good.

Detachment training in the spring of 1967 now concentrated on the prospects of Project OXCART going from operational training to actual operational. The flights were routine flying for pilot proficiency and systems checkout, however, emphasizing two aerial refueling missions. Such profiles were like those proposed for operational missions in the event the CIA activated the SKYLARK or BLACK SHIELD contingencies. Construction at Kadena was near completion, except for new hangars, and with provisions made to use other hangar facilities on a temporary basis should the need arise. In a crisis, the CIA could extend the deployment route to include photographic coverage of North Vietnam with recovery at Kadena,

Kadena, Okinawa

OXCART's BLACK SHIELD Deployment, Operations, and Termination.

Finally Extending Area 51's Power Projection Activities to Kadena, Okinawa
Finally, after project Area 51 was declared operational, the CIA and its support elements were taking the birds out of their nest and letting them fly.

Kadena Support

Before deploying the planes and personnel from Area 51, the OXCART honcho, Warner Weiss prepositioned 1,000,000 pounds of equipment at Kadena Air Base. He hired contractors on the island to construct an operations building, hangars. He placed 19 people in place to prepare the POL fuel storage and to support the CIA's equipment and facilities.
CIA's Werner Weiss made several trips between Area 51 and Kadena, Okinawa to arrange billets for the OXCART personnel placed on temporary duty at Kadena for Operation BLACK SHIELD. He arranged for sheltering the ultra-secret A-12s and the equally classified air force and civilians supporting BLACK SHIELD. He did a Herculean job of converting Quonset buildings into living quarters a Morgan Manor, the housing area occupied by the CIA, contractors, and 1129th SAS in 1967 to 1968.

Morgan Manor

Morgan Manor was a group of WW-2 Quonset Huts in a little compound that was a mile north of the small town of Kadena Circle, at the north edge of the air base. It appeared to have been a US dependent housing area. Each Quonset hut divided in half with an entrance at each end. Each half had a living room, kitchen, bathroom, and four bedrooms. The closets in the bedrooms had two, high wattage light bulbs, near the floor, which was on continually, to try to reduce the humidity and to prevent mold.

There were a laundry and cleaners; both were in mobile home type trailers at the entrance to the compound and across from our mess. An old, yellow dog hung around the Mess trailer, probably for scraps. The dog answered to the name of, "Hi, Dozo" [Japanese for: 'Yes, Please']. House Maids gathered bundled the laundry for those living there and for security, Okinawan guards dressed in gray uniforms with white pith helmets patrolled Morgan Manor unarmed except for nightsticks.

Morgan Manor was adjacent to Kadena Air Base. Kadena Circle was at the junction of North/South Highway One and the Cross-Island Highway at the edge of the Air Base. The town was the favorite assembly point of the "Okinawa Reversionist Movement" (on Sundays). Their purpose was to get Okinawa to "revert" to Japan, and one of their favorite chants was "B-52 Go Home!" The BLACK SHIELD participants had strict orders to avoid the demonstrations at all cost.

There were two bus systems on Okinawa, The Ryukyu Bus Co., and The Okinawa Bus Co. Some of their routes overlapped since the main highways run along the edges of the Island. The two companies were as different as night and day. The rust color Ryukyu buses appeared to had rust spots, they rattled, smoked, and were in general disrepair. The Okinawa buses, painted a cream color, with light green trim, all seemed in good shape. The drivers decorated the inside of each bus with plastic flowers and played the radio for the passengers. When an American came onboard, the driver changed stations to get US music.

A monument for Ernie Pyle stood in the middle of a tree-ringed plot with a very moving tribute on a brass plaque. It reads, "ON THIS SPOT THE 7th INFANTRY LOST A BUDDIE ERNIE PYLE, 17 APRIL 1945." Fields of the island farmers surrounded the plot where he was buried on site with other GI casualties. (The plots later moved to the US Army cemetery on Okinawa.)

The BLACK SHIELD personnel sympathized with the locals' desire to get rid of the B-52s, that the GIs called BUFs (Big Ugly Fellows). The Vietnam War was near its peak, so there were nearly daily bombing raids flown out of Kadena. The Okinawa "Morning Star" paper said that they were bombing the North Viet's supply base in the Ashua Valley. The range of the targets was so great that it was necessary to refuel the bombers in flight. The B-52s took off between 3 and 5 A.M to get the B-52s over the targets during daylight. The Tankers took off 1 to 2 hours earlier, so you could expect to hear the continuous roar of heavily loaded airplanes from about midnight until 6 A.M. Then some of them would start returning about 7 A.M. The poor people who lived near the air base could not get a good night's sleep! The sounds of the B-52s taking off was nothing compared to the sounds of the A-12 or SR-71 taking off.

Headquarters in Washington commanded and controlled OXCART operations at Kadena to enable flying operational missions ten days after mission approval.

The OXCART program supported nine operational missions per month using the three aircraft deployed to Kadena for Operation BLACK SHIELD.

CIA housing compound during Operation BLACK SHIELD at Kadena, Okinawa in 1967 – 1968 – Roadrunner Historian Frank Murray

Controlling the Activities

From the start of the OXCART project, there existed an informal agreement that required the air force to give the CIA complete coordination of all its elements. This had applied to OXCART during the flight test and training at Area 51, and then the operational flight training carried out by syllabi and standards as mutually agreed between the CIA and the air force. In 1965, the flight testing had ended, and operational flight training began. OXCART was now ready to go operational.

Going operational became a possibility during May 1967, over a growing concern over the possible introduction of the enemy offensive surface to surface missiles into North Vietnam.

The CIA discussed the problem of detecting such missiles with the president on 11 May at which time he requested a proposal on how best to monitor the situation.

The CIA responded the next day with a briefing to the 303 Committee on the photographic average required for missile search and the extent of existing coverage, which was inadequate. The CIA presented a proposal to use the A-12s.

The State and Defense members of the 303 Committee wanted to examine the political risks of the requirement further. While they conducted their examination, on 15 May 1967, Mr. Helms submitted to the 303 Committee a formal proposal to deploy the OXCART.

The following day, Mr. Rostow reported the president had given his approval for immediate deployment and use of the OXCART over North Vietnam. The CIA immediately placed the BLACK SHIELD contingency plan into effect.

On 17, 18 and 19 May, the CIA and the 1129th SAS accomplished a logistical airlift to Kadena. Six C 141 aircraft transported approximately. 120,000 tons of cargo and equipment and the task force personnel from Area 51. The CIA positioned communications and support teams at Hickam AB and at Wake Island to assist in the recovery of the aircraft in case of emergency landings at either site. The CIA also confirmed arrangements to brief the relevant ambassadors fully and to advise the Prime Ministers of Japan and Thailand, the president and defense minister of the Republic of China and the Chiefs of the air force of Thailand and the Republic of China. The CIA received favorable reactions from all.

On the fifth day after receiving deployments orders for the CIA headquarters, the first A-12 departed and traveled the 6,673 miles in five hours and 34 minutes. The second plane left on the seventh and the third on tje ninth day.

Washington had arranged for someone to brief the Ambassadors and Chiefs of Station in the Philippines, Formosa, Thailand, South Vietnam, and Japan, and the High Commissioner and Chief of Station, Okinawa. The Prime Ministers of Japan and Thailand received notice as did the President and Defense Minister of the Republic of China. The chiefs of the Thailand and the Republic of China's air forces received briefings with favorable reactions.

The first two became ready for an emergency mission on the eleventh day, and for a typical mission on the fifteenth day.

On 29 May 1967, the unit at Kadena stood prepared to fly an operational mission. Under the command of Colonel Hugh C. Slater, two hundred and sixty personnel deployed from Groom Lake to the BLACK SHIELD facility in Okinawa. Except for hangars a month short of completion, everything appeared in shape for sustained operations. The very next day, Washington alerted the detachment for a mission on 31 May, the date that all concerned saw the culmination of ten years of effort, worry, and cost.

On 22 May, the first A-12 Article 131, flown by Mele Vojvodich, flew to Kadena from Area 51, non-stop in 6:06 hours for 6,874 nautical miles with refueling west of San Francisco, near Hawaii and as it neared Wake Island. There were no significant aircraft malfunctions, and the flight was completed as planned without difficulty.

Article 127, Flown by Jack Layton, departed Area 51 on 24 May and arrived at Kadena 5:55 hours later. Again, the CIA pilot accomplished the flight, including three refueling, without difficulty. The A-12 attained an average speed of Mach 2.03 from take-off to landing that included refueling time. After landing at Kadena, Layton mentioned his plan position indicator (PPI) display not working in his A-12. When asked with he noticed the malfunction; he confessed that it was not working with he took off at Area 51.

The third A-12, Article 129, flown by Jack Weeks, launched as planned on 26 May 1967. The flight proceeded normally until the pilot experienced an inertial navigation system (INS) problem and communication difficulties near Wake Island. Under the circumstances, he made a precautionary landing at Wake Island.

The CIA had provided for such a contingency by prepositioning an emergency recovery team who secured the aircraft without incident. Article 129 continued its flight to Kadena the next day.

On 29 May 1967, the unit at Kadena stood prepared to fly an operational mission. Under the command of Colonel Hugh C. Slater, two hundred and sixty personnel deployed from Groom Lake to the BLACK SHIELD facility in Okinawa. Except for hangars a month short of completion, everything appeared in shape for sustained operations. The very next day, Washington alerted the detachment for a mission on 31 May, the date that all concerned saw the culmination of ten years of effort, worry, and cost.

Command, Control, and Communications.

Targeting, flight planning, and command of the OXCART vehicle centered at CIA headquarters in Washington, DC. CIA headquarters also prepared the flight plans that it transmitted to Area 51 and Kadena via a 1004 high-speed, secure digital data circuit. The ground facilities and tanker aircraft coordinated through high-frequency single sideband radio, UHF radio links, KW-26 secure teletype circuits, and both secure and hotline telephones. A high-frequency BIRDWATCHER system monitored the A-12 in flight with the capability of recall when necessary.

Mission preparation time allowed for aircraft, sensor and crew generation and required 24 hours, or sooner for canned missions pre-planned, and with the planes and crews in the countdown stage.

The final decision as to execution and timing of actual overflight missions rested jointly with the CIA and the air force. The decision was subject to such guidance as received from higher authority, and by notification, coordination, and support procedures currently employed in project OILSTONE. The line of command was direct between operational units and the CIA. The responsibility for the overall security of the program rest with CIA. Given the security aspects of this project, the CIA and air force headquarters considered it necessary that the project had maximum practicable compartmentation that included provisions for logical, innocent explanation of the activities.

The CIA forwarded the notification of an impending mission to the detachment by Secure Communications, this usually occurring 24 hours in advance of the mission take-off time. Now, the detachment Commander selected the primary and backup pilot for the mission. They picked the primary pilot on a rotation basis, him the backup pilot on the previous mission. Both pilots attended the mission briefings.

On the morning of the first operational mission, torrential rains fell at Kadena. The weather over the target area remained clear, so preparations continued in hopes of the local weather clearing. The time for takeoff approached. Until now, the A-12 had never operated in torrential rains. CIA pilot Mele Vojvodich taxied to the runway and took off while the rain continued.

Vojvodich flew the first operational overflight of North Vietnam by an A-12. The mission launched in a rainstorm, probably the first time for an A-12. Mele Vojvodich's mission was a success, and the Intelligence world got their first taste of the high-quality photography from the Perkin-Elmer type 2-camera system

The route of the first BLACK SHIELD mission followed one flight line over North Vietnam and one over the Demilitarized Zone. The mission lasted three hours and 39 minutes, and the cruise legs flew at Mach 3.1 and 80,000 feet with satisfactory results. The A-12 photographed seventy of the 190 known SAM sites in North Vietnam along with nine other priority targets. The A-12 detected no radar signals, indicating that the first mission completely unnoticed by both Chinese and North Vietnamese.

Washington alerted fifteen BLACK SHIELD missions during the period from 31 May to 15 August 1967. All the missions were planned, directed, and controlled by project headquarters against targets identified by the United States Intelligence Board (USIB) made up of the top officials of large organizations and approved by higher authority.

The CIA flew seven of the fifteen and the seven, four detected radar tracking signals. By mid-July, they had determined with a high degree of confidence that no surface-to-surface missiles existed in North Vietnam. (Note. North Vietnam proved to be a difficult region for flying Mach 3 reconnaissance flights because of the narrowness of his country. At speed, the A-12 required up to a 125-mile radius to make a turnaround, making it difficult to avoid unintended penetration across borders.)

They quickly established a routine of flying each of the A-12s once per week. The CIA pilots flew for some reasons, an operational mission or to exercise the airplane and give currency training to the pilots. Most of the time they received enough operational missions to provide the exercise needed. The pilots rotated back to Area 51 at a six-week interval, but at times, made changes as needed.

In June, with the onset of the monsoon, the entire North Vietnam area promised less than three days of weather suitable for photography. From October 1965 to April 23, 1966, out of 87 missions flown, less than a half of the CIA's priority objectives had produced baseline data because of the weather. Unlike the satellites, OXCART could program coverage when there were sudden openings during periods of sustained bad weather — and the weather was a prime consideration in South China and North Vietnam. There may be one or two days a month when a large hole or "bubble" would open in the overcast and permit photography. These bubbles were relatively predictable to the weathermen at 24 intervals ad it was blown across China from west to east. The satellites could not be programmed to take advantage of this bubble. However, a manned aircraft could use these openings to gather database data and changes from the base on the Chinese military posture, which the Chinese resorted to diversionary tactics to hide.

Consequent, at Langley, the OXCART mission planners maintained a constant watch on the weather in the target areas. At 1600 hours each day project headquarters held a mission alert briefing, and if the forecast weather appeared favorable, the headquarters alerted the field to a mission and provided a flight route, both preceding an actual mission take off by 28 to 30 hours. The mission planners conducted a second review of target weather twelve hours (H 12) before taking off. The mission generation sequence continued if the weather continued favorable. At H 2 hours, the CIA made a "go nor go" decision and communicated it to the field.

The final decision, it should be noted, was not just based on the target area weather factor. The weather had to be adequate also in the refueling areas and at the launch and recovery base for the safety of flight, reasons.

In the field, operations and maintenance generation began with the receipt of alert notification. Headquarters selected a primary aircraft, and pilot, as well as a backup aircraft and pilot. At Kadena, the aircraft received a thorough pre-mission inspection and servicing, checking all systems, and loading the camera into the plane. The pilots did not get a detailed mission route briefing until in the early evening before the day of the flight.

On the morning of the flight, a final briefing reported the aircraft and system status and briefed the pilot and the backup pilot on last-minute weather, intelligence, and any flight plan amendments or changes. Two hours before taking off, the primary pilot received a medical examination, was suited, and transported to the aircraft. In the event of equipment malfunction or for other reasons, the primary aircraft could not take off, the spare or backup was prepared to execute the mission one hour later,

A typical route profile for a BLACK SHIELD mission over North Vietnam included a refueling shortly after taking off, south of Okinawa, accomplishing planned photographic pass, withdrawing to a second aerial refueling in the Thailand area, and returning to Kadena,

Mission timing, tactics and routes varied consistently with photographic requirements and threat analyses,

The 1129th SAS weather team at Kadena during Operation BLACK SHIELD. Note the Roadrunner logo. CIA via TD Barnes Collection

Unlike the Air Force, the CIA did not have unit designations or shoulder patches. The Air Force support felt they needed a patch, so they designed the Roadrunner patch below that various sections altered to fit them. The A-12 pilots flying during Operation BLACK SHIELD wanted their own patch, so Jack Weeks designed the Cygnus constellation symbol also became the subject of a crew patch worn by the CIA A-12 pilots. Though neither patch was official, they are both considered a part of the CIA legacy at 'Area 51. The participants in the CIA U-1 and A-12 projects now have a social association called Roadrunners Internationale. The associations has expanded to include the YF-12 and MD-21 participants at Area 51 during that era.

A-12s and Pilot Assets.

After the establishment of the BLACK SHIELD capability at Kadena Air Base, Okinawa in early 1967, the CIA positioned three A-12s and a support detachment to maintain and operate the A-12s for the Recon Mission in the Eastern Pacific Region. There were plans to have two A-12 pilots on station for missions always. During the times when a replacement pilot was in place, there were three pilots for a week or so. The original three A-12s remained in place for the duration of BLACK SHIELD because the BLACK SHIELD Operation ended before the plans to rotate the A-12s from Area 51 could occur.

The pilots and the Area 51 command staff rotated on a six-week interval, making scheduling changes as required.

At the time of the BLACK SHIELD deployment, Project OXCART provided six qualified operational A-12 pilots. The activation of the detachment at Kadena started in May 1967 with the move of three A-12s from Area 51 to Kadena.

Mele Vojvodich, Jack Layton, and Jack Weeks were the first project pilots to rotate on TDY, a temporary duty from Area 51 to Kadena. Frank Murray, Ken Collins, and Dennis Sullivan remained at Area 51 where they continued enough training flights to maintain their currency.

The first A-12 made the trip without a problem, flown by Mele Vojvodich, another A-12, flown by Jack Layton made the trip easily. The last A-12, flown by Jack Weeks had some difficulties with an inertial navigation system and Comm, and he made a stopover at Wake Island, followed by a move on to Kadena the next day.

Chapter 11 – BLACK SHIELD Missions

Mele Vojvodich, Jack Layton, and Jack Weeks were the first project pilots to rotate from Area 51 to Kadena. Frank Murray, Ken Collins, and Dennis Sullivan remained at Area 51 where they continued enough training flights to maintain their currency.

BSX001

On 31 May 1967, Mele Vojvodich, Dutch 30, flew A-12 Article 131 on BLACK SHIELD Mission BSX001, a single-pass reconnaissance mission that entered North Vietnam over Haiphong at 0313:46 hours ZULU and exited over the demilitarized zone (DM hours ZULU) at 0421:45 hours ZULU.

The Vojvodich mission photography gave no evidence of offensive surface-to-surface missile equipment (SSM) or facilities. The CIA knew of 190 North Vietnamese SA-2 sites and photographed 70 of them along with the Chinese Naval Base at Yu-lin. Vojvodich's mission covered nine of the 27 COMIREX priority targets in North Vietnam.

He saw no sign of Chinese or North Vietnamese tracking. His SIP ELINT collection system (A SIGINT and IMINT Processor) did not record any FAN SONG signals, and he saw no sign of a weapons reaction while over hostile territory. The Department of Defense (DOD) strike jamming operations during the mission flight period stayed light, with no EB-66B or EB-66C jamming aircraft operating during the reconnaissance mission.

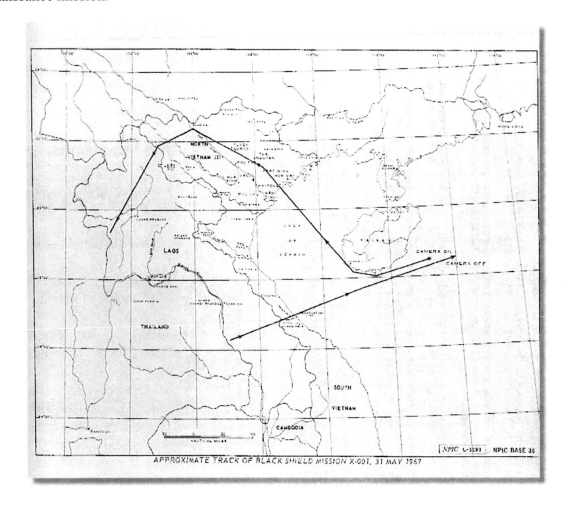

BSX003

On 10 June 1967, Jack Weeks, Dutch 29, flew A-12 Article 131 on BLACK SHIELD MISSION BSX003, a single-pass reconnaissance mission flown over North Vietnam. The mission aircraft entered I North Vietnam over Haiphong.

The Weeks mission's photography showed no evidence of offensive surface-to-surface equipment or facilities. The CIA knew of 193 North Vietnamese SA-2 sites before the mission and photographed 97 of them during the BSX003 mission. The 97 sites included four previously unidentified missile sites and 18 occupied SA-2 sites. Weeks covered 13 of the 27 COMIREX Priority l targets in North Vietnam with no indication of tracking or a weapons reaction by either the Chinese or the North Vietnamese. His SIP ELINT collection system did not record any FAN SONG signals. The DOD operational strike jamming activity stayed heavy during the overflight period, with four EB-66C and one EB-66B aircraft conducting barrage and spot jamming at the time. The jamming affected many early warning and fire control radars.

BX6705

On 20 June 1967, Jack Layton, Dutch 27, flew A-12 Article 129 on BLACK SHIELD Mission BX6705, a double-pass reconnaissance mission flown over North Vietnam. Layton's A-12 penetrated and exited over the DM hours ZULU.

The Layton mission's photography gave no evidence of surface-to-surface missile equipment or facilities in North Vietnam. The CIA knew of 200 North Vietnamese SA-2 sites before the mission. BX6705 photographed 133 of them including two previously unidentified sites. The mission found 20 of the sites occupied and covered 17 of the 27 COMIREX Priority I targets in North Vietnam.

The mission aircraft saw no known hostile weapons reactions to the mission, nor did Layton's SIP ELINT collection system record any FAN SONG signals. However, Layton did exercise the Big Blast manual jam override while near Hainan Island on the return leg of the route. The Big Blast, an onboard defense mechanism transmitted S and C-band noise energy for 90 seconds.

The first known tracking of BLACK SHIELD aircraft occurred on this mission. A communist Chinese Chiang Cheng radar facility reported the A-12 as it passed eight position points and eight altitudes. The first detection range and the final radar vehicle distances were respectively 86 nautical miles and 82 nautical miles. This radar station operated by a MOON FACE radar and a ROCK CAKE radar.

The Chinese Chiang Cheng radar reported jamming from an unknown source on both its MOON FACE radar and its ROCK CAKE height-finder radar.

The jamming activity probably influenced the ROCK CAKE radar operator and in searching these azimuths to identify the originator of the activity, detected and then tracked the mission aircraft. The fact that the radar reported altitudes with each valid plot position supported the tentative conclusion that the Chiang Cheng ROCK CAKE height-finder radar tracked the mission.

No BLACK SHIELD aircraft jammers operated during this period, or any DOD operational reports of a jamming activity. His SIP ELINT collection system was inoperable during the tracking period, and thus it was impossible to correlate the tracking data from the radar stations and the ELINT data.

The radar facility also rebroadcast the track information to Kun-ming some 14 hours after the mission flight. There was no indication of North Vietnamese tracking or the passage of tracking data to the Vietnamese. DOD strike activity was very light during the mission overflight.

BX6706

On 30 June 1967, Jack Weeks, Dutch 29, flew A-12 Article 129 on BLACK SHIELD Mission BX6706, a double-pass reconnaissance mission flown over North Vietnam. Weeks entered Vietnam north of Haiphong and exited over the DM hours ZULU.

The Weeks mission's photography gave no evidence of surface-to-surface missile equipment or facilities. The CIA knew of 205 North Vietnamese SA-2 sites before the mission and photographed 109 of them including three previously unidentified sites. The mission found 16 of the sites occupied. Week's mission covered 21 of the 27 COMIREX Priority I targets in North Vietnam with no sign of a hostile weapon reaction. On this mission, his System 6 ELINT collection package that replaced the SIP, SIGINT and IMINT Processor, recorded six FAN SONG signals during the mission overflight. None of these signals suggested their tracking the mission aircraft, nor were any defensive systems activated because of this activity. Haikou, a Chinese air defense zone facility on Hainan Island reported mission track information thirteen and a half hours late.

The unidentified radar station(s) did not report any tracking altitudes. Considering the region of the tracking and the fact that all radar tracks on Hainan Island reported information to Haikou for rebroadcast, the CIA doubted the location of the tracking radar being on Hainan Island. There's no indication that anyone passed this track information to the North Vietnamese.

An EA3B ELINT collection platform in orbit during the overflight period did intercept ROCK CAKE signals emanating from the Huang-Hu area of Hainan Island. The period and duration of the intercept signals compared favorably with those of the track period. The signal might have been the radar station tracking the mission aircraft.

In the period from 0530 hours ZULU to 05402, his System 6 ELINT collection tapes recorded a probable BIG MESH V-beam S-band radar changing scan modes from circular to steady to circular. The signal indicated radar operator interest in the mission aircraft or interest in the volume of space in which the mission aircraft happened to be flying. COMINT provided no sign of North Vietnamese tracking during this period. No known DOD strike jamming operations were active during the mission overflight.

BX6708

On 13 July 1967, Kenneth Collins, Dutch 21, flew A-12 Article 127·Mission BX6708, a single-pass, high-altitude reconnaissance mission flown over North Vietnam, entering south of Haiphong.

Collin's Mission photography did not show any evidence of surface-to-surface missile equipment or facilities. The CIA knew of 214 North Vietnamese SA-2 sites, and Collins photographed 71 of them including one previously unidentified site. Of the 71 sites covered, the mission found five occupied and covered eight of the 27 Priority 1 targets in North Vietnam.

Collin's mission saw no sign of a hostile weapons reaction. His System 6 ELINT collection package recorded four FAN SONG signals during this mission. However, these signals did not appear to be tracking the mission aircraft, and none of the A-12's defensive systems activated because of these signals. He saw no COMINT intercept evidence of mission tracking by either Chinese or North Vietnamese radar facilities. Analysis of System 6 records indicated that two probable BIG MESH, V-beam, S-band radars changed scan modes from circular to manual to circular. Although these changes in scan mode might have shown an interest in the same volume in space as the A-12, no COMINT evidence suggested tracking by either the North Vietnamese or the Chinese.

There was little or no DOD strike jamming activity while Collins' aircraft was over North Vietnam. One EB-66C, an ELINT collection active jamming platform, was orbiting the area during the mission overflight period. Overall signal density was light.

BX6709

On 19 July 1967, Dennis Sullivan, Dutch 23, flew A-12 Article 131 on BLACK SHIELD Mission BX6709, a double-pass, high-altitude reconnaissance mission flown over North Vietnam. The CIA knew of 2 hours ZULU15 SA-2 sites in North Vietnam, and Sullivan photographed 166 of them on this mission, including two previously unidentified sites. Of the 166 sites photographed, his mission found 12 occupied and covered 12 of the 27 COMIREX Priority I targets.

There was no sign of a hostile weapons reaction His System 6 ELINT collection package recorded nine FAN SONG radar signals during this one mission. These signals ·did not appear to be ·tracking the mission aircraft. None of the defensive systems activated because of these signals. Chiang Cheng, a Chinese radar station reflected reporting track confirmation on the mission aircraft in seven minutes of continuous tracking. The track plot was compatible with the mission flight route, but reported altitudes ranged from 8,000 to 11,000 feet higher than the actual mission aircraft altitude. The CIA never identified the type of radar generating these errors that China shared this track information with the North Vietnamese. DOD strike jamming activity was moderately heavy during the mission overflight period. United States ECM aircraft in operation in the area during the Sullivan mission overflight period were: two EA3Bs, three EA1Fs, one EB-66C, and one EB-66B.

BX6710

On 20 July 1967, Kenneth Collins, Dutch 21, flew A-12 Article 129 on BLACK SHIELD Mission BX6710 on a double-pass, high-altitude reconnaissance mission flown over North Vietnam. Collins' A-12 aircraft entered North Vietnam north of Haiphong and exited over the demilitarized zone. Collins' mission photography gave no evidence of surface-to-surface missile equipment or facilities. The CIA knew of 219 SA-2 sites in North Vietnam and photographed 80 of them on this mission including one previously unidentified site. Of 80 of the sites covered, Collins found five of them occupied and covered 19 of the 27 COMIREX Priority 1targets in North Vietnam.

A reassessment of North Vietnam's rail inventory from the photography of back-to-back missions BX6709 and BX6710 showed the Vietnamese having a rail inventory three times larger than previously believed. By this photography, North Vietnam had an estimated 2, 000 to 3, 000 freight cars and 100 to 120 locomotives. Photography of the P'ing-hsiang China transshipment point from missions BX6709 and BX6710 showed 100 to 130 field artillery pieces of unknown caliber on a nearby railroad siding. These artillery pieces were enough to equip two to three regiments for use in North Vietnam.

Collins saw no sign of a hostile weapons reaction, nor did his System 6 ELINT collection tapes record any FAN SONG signals during this overflight. There were two other high-altitude reconnaissance missions operational during the BX6710 overflight, Church Door C237C and Blue Springs BSQ-19P-V638.

The enemy tracked Mission C237C continuously while in the South China area and BSQ 190 continuously while over North Vietnam and the Gulf of Tonkin. Chinese air defense facilities reported track plot times on BX6710 and reported altitudes compatible with those of the BLACK SHIELD aircraft. Two Chinese radar stations reported track information.

Collins' BX6710 mission path crossed the path of the continuously tracked C237 C flight and the Chinese broadcast facilities reported an initial plot point of the BX6710 vehicle. A similar phenomenon occurred over the Gulf of Tonkin where the Chinese Pei-li radar station reported track information on Blue Springs Mission BSQ-190. At 0507 hours ZULU, BX6710 entered the radar region of the Blue Springs vehicle tracking. The Chinese facilities reported the initial plot point of the second pass as occurring at 0507 hours ZULU time. It appeared at least in these two instances that a path crossing with an identified tracked vehicle enhanced the initial detection capability of the radar operator.

In the past, initial radar detection occurred at about the time the radar station had the benefit of the vehicle's broadside radar cross-section. Hanoi Bae Mai reported track plot times of 0455 hours ZULU to

0511 hours ZULU on the preceding vehicle of the mission C237C. The Chinese passed BX6710 track plot time data of 0507 hours ZULU to 0518 hours ZULU to the North Vietnamese via the Kuang-chou-DRV liaison link. Hanoi Bac Mai reported BX6710 track plot time data of 0511 hours ZULU to 0518 hours ZULU. It seemed unlikely, considering the sequence of events, that the Hanoi Bae Mai hostile broadcast reflected Vietnamese radar tracking, but rather that the broadcast was a rebroadcast of Chinese BX6710 tracking. In any case, this was the first known instance of actual North Vietnamese knowledge of a mission vehicle flight. DOD strike jamming operations were light during the mission overflight period.

BX6716

On 21 August 1967, Mele Vojvodich, DUTCH 30 flew A-12 Article 131 on BLACK SHIELD Mission BX6716, a double-pass, high-altitude reconnaissance mission flown over ·North Vietnam. His A-12 reconnaissance vehicle entered and exited North Vietnam on its first pass at 0504:27 hours and 0515:36 hours ZULU and its second pass at 0619:14 hours and 0628:28 hours ZULU.

Vojvodich's mission Photography gave no evidence of offensive surface-to-surface missile (SSM) equipment or facilities. Clear photograph covered Hanoi. However, some clouds and haze over North Vietnam and China limited the ability to interpret photography of Haiphong and some others. Vojvodich photographed 156 COMIREX targets plus five additional bonus targets. Of these, 108 were SA-2 sites, including one previously unknown site. The Vojvodich mission found fourteen of the photographed SA-2 sites occupied.

Vojvodich's mission provided good bomb-damage assessment photography of the Hanoi electric power plant, enabling analysts to estimate probabilities: damage to two turbine generators· Photographic coverage of 18 bridges and 11 rail yards yielded current information on the status of the transportation.

Vojvodich saw no indication of a hostile weapon reaction directed against his A-12 reconnaissance vehicle. However, onboard warning defensive systems did activate during the first pass of the overflight at 0506:34 hours ZULU. The following systems on Vojvodich's mission noted activity indicative of a missile launch and subsequent guidance:

(a) The System 6 onboard ELINT collection device,
(b) Blue Dog, an L-band guidance jammer,
(c) Big Blast, an S- and C-band noise jammer, and
(d) The missile launch indicator lights.

However, analysis of his System 6 ELINT collection tapes tape indicated that none of the seven Fan Song signals collected appeared to be tracking the mission vehicle. His System 6 tapes showed the L-band guidance signal coming up before Vojvodich's arrival in the Fan Song sector. The Fan Song was in high PRF.

The 20-second duration of the L-band signal indicated a missile flight to an altitude of 15,000 to 20.000 feet, an altitude compatible with the operational tactics of some of the DOD strike aircraft operating in North Vietnam.

An analysis of the Blue Dog tapes indicated the missile never fired on a path t0 intercept Vojvodich's A-12. COMINT did not indicate Fan Song tracking or a missile launch towards the Vojvodich vehicle. However, COMINT did show several missiles fired at some Strike aircraft operating in the Kep and Lang Son area during this period. Thus, it seemed most likely the radar missiles activated the onboard systems were directed at the strike aircraft and not the CIA's A-12.

None of the onboard systems activated while over North Vietnam during the second pass. The Chinese air defense carried out air surveillance tracking. However, the CIA found no indication of the Chinese passing on the track information the North Vietnamese. The tracking of the mission vehicle was limited.

Possibly because of the intense DOD strike jamming operations occurring during the overflight period. Five EB-66B, two EB-66C, arid one EA-lF aircraft actively engaged in the jamming and the overall heavy signal density.

BX6718

On 31 August 1967, Jack Layton, Dutch 27, flew A-12 Article 127 on BLACK SHIELD Mission BX6718 on a double-pass reconnaissance mission flown over North Vietnam. Layton's A-12 reconnaissance vehicle entered North Vietnam on its first pass at 0402:22 hours and exited on its final pass at 0645:46 hours ZULU.

Layton's photography gave no evidence of surface-to-surface missile equipment, or facility camera malfunctions occurred on both the first and second passes. A failure in the inertial navigation system (INS) also took place on the second pass, resulting in Layton photographing only 13 COMIREX targets, three of them unoccupied SA-2 sites.

Layton saw no indication of a weapons reaction to the mission vehicle. The System 6 tapes recorded eight Fan Song signals during the overflight period. However, none of the sites appeared to be tracking Layton's A-12, and no onboard defensive systems activated because of these signals.

Chinese air surveillance facilities tracked the mission vehicle, and two Chinese radar stations reported the track information. The plot from Ho-lo-Shih began at 0347 hours ZULU with the aircraft at a ground range of 185 nautical miles. Moreover, ended at a radar-to-aircraft ground range of 140 nautical miles. Chiang-cheng's initial plot occurred at a ground range of 134 nautical miles with the final plot occurring at a radar-to-aircraft ground range of 215 nautical miles. The Kuang-chou Sino-North Vietnamese liaison facility passing track information on Layton's plane to the North Vietnamese, a rebroadcast of the same information occurred 11 minutes later. This reporting time delay enabled Layton to complete the first pass over the SA-2 threat zones before the initial air surveillance early warning.

Very light strike jamming activity by US aircraft occurred during the overflight period with one EB-66C and one EB-66B operating in the vicinity from 040:2 to 0505 hours ZULU.

BX6722

On 16 September 1967, Jack Weeks, Dutch 29 flew A-12 Article 129 on BLACK SHIELD Mission BX6722, a double-pass, high-altitude reconnaissance mission flown over North Vietnam. Weeks entered and exited North Vietnam on his first pass at 0412:57 hours and 0423:52 hours ZULU and on its second pass at 0524:04 hours and 0534:20 hours ZULU.

He found the weather in the target area much worse than anticipated, with most of the target area under heavy cloud cover. He photographed 45 COMIREX targets in the North Vietnam-China area. Seventeen of these objectives were Vietnamese surface-to-air missile sites, four of them occupied. Weeks' photography gave no evidence of surface-to-surface missile equipment or facilities. He saw no indication of a hostile weapons reaction. His System 6 recorded two Fan Song signals during this mission, but these signals did not appear to be tracking the mission aircraft, and no defensive system activated by these signals. Chinese air surveillance defense facilities reported track plot times on this mission at a radar-station-to-aircraft ground range of 128 nautical miles and ended at 0418 hours ZULU at a ground range of 105 nautical miles.

The first occurred on BLACK SHIELD mission BX6710 flown on 20 July 1967 as noted. The initial plot from Chiang-cheng occurred at a radar-to-aircraft ground range of 140 nautical miles, and the final plot took place at a ground range of 160 nautical miles. North Vietnam's host. The broadcast facility Hanoi was again noted reflecting the Chines air surveillance plot information of the BLACK SHIELD vehicle. One EB-66B and one EB-66C active ECM platforms were in operation over the Gulf of Tonkin from 0254 hours to 0502 hours ZULU. •

BX6723

On 17 September 1967, Kenneth Collins, Dutch 21 flew A-12 Article 131 on BLACK SHIELD mission BX6723, a double-pass, high-altitude reconnaissance mission flown over North Vietnam.

Clear photography covered Hanoi, Haiphong, and the Chinese border area and only about 10 percent of the target area covered with clouds. Collin's photography gave no evidence of surface-to-surface missile facilities or equipment. Even though the Soviet freighter Partizanskaya Iskra, an alleged carrier of SS-N-2 Styx cruise missiles, was noted berthed and with all five hatches open. A search of the ship to port area failed to detect any SS-N 2 crates or equipment. Mission photography also provided bomb damage assessment of the port areas and land transportation network. Collins photographed over 200 COMIREX targets of North Vietnam and China. Approximately 150 of these were North Vietnamese SA-2 sites, including four newly identified sites, nineteen of them occupied.

Collins saw no indication of a hostile weapons reaction. However, North Vietnamese surface-to-air missile communications did reveal a hostile intent when a surface-to-air missile site went to a "condition one." "Condition one" in Soviet-Vietnamese terminology defined the highest readiness state of a surface-to-air missile site, COMINT indicated that a second site tracked the vehicle with its acquisition radar but was unsuccessful with its Fan Song guidance radar.

The North Vietnamese SAM communication made references to a very fast speed and altitude of 25,000 meters. COMINT indicated that a second site was unsuccessful in its attempt to pick up the vehicle on its guidance or acquisition radars. The first mention of the BLACKSHIELD aircraft by the North Vietnamese SAM air defense system occurred on mission BX6723. Collin's System 6 recorded five Fan Song signals during the overflight, none of them appearing as emanating from sites tracking his A-12. His defensive systems never activated because of these signals.

Collin completed his mission with complete, unbroken air surveillance tracking his A-12 while over North Vietnam. Three Chinese radar stations of the Chinese air defense system carried out this air surveillance tracking and passed it to the North Vietnamese. The Chiang-cheng radar station's initial detection of the vehicle occurred at 0513 hours ZULU at a radar-to-aircraft ground range of 108 nautical miles, and the final plot, at 0517 and a ground range of 218 nautical miles. Collins saw no indication of any DOD strike jamming activity during the overflight period.

BX6725

On 4 October 1967, Kenneth Collins, Dutch 21, flew A-12 Article 127 on BLACK SHIELD Mission BX6725 on a high-altitude, double-pass reconnaissance mission flown over North Vietnam. Collins intercepted two Fan Song signals on the first pass, which activated the onboard defensive systems, but saw no indication of a missile launch. He did not intercept any threat signals during the second pass.

Collins' mission BX6725 obtained clear photography over 40 percent of the country, including the Hanoi, Haiphong, and Thai Nguyen areas and the key Ping-hsiang and Dong Dang transshipment points on the Chinese-North Vietnamese border. Mission photography revealed for the first time the presence of armored vehicles on flatcars at Ping-hsiang, China, where combat equipment was limited heretofore to field artillery and AA guns. His photography gave no evidence of surface-to-surface missile deployment. He photographed 18 COMIREX and 16 bonus targets, 154 of them SA-2 sites, including 18 occupied sites.

At 0335 hours ZULU, the Kuang-chou District broadcast the initial detection of Collins' A-12 with a first plot report time of 0333 hours ZULU. The Hanoi Bae Mai hostile broadcast, upon reception of the Kuang-chou transmission of 0335 hours ZULU, began tracking Collins in the A-12. At 0338 hours ZULU, a surface-to-air missile battalion alerted the senior regimental surface-to-air missile controller to the approach of a high-altitude, very fast target at a range of 108 nautical miles. This detection range proved compatible with the ·Spoon ·Rest A radar that collocated with the battalion. Shortly after entry into North

Vietnam, Collins intercepted Fan Song signals, which in turn activated his Mad Moth jammer.

According to COMINT, two surface-to-air missile battalions located in the Hanoi area were actively tracking Collins' mission during the period. A message intercepted two days later 6 October, reflected the surface-to-air missile controller at site VN 142 conversing with his regimental controller. The former stated, "Several units picked up the target, the SR-71 . . . We could not pick it up because it was too fast Those that picked it up lost it within 3 minutes."

These statements of 6 October coincide with the events of 4 October, Pin Peg warning lights located in the cockpit also indicated threat activity emanating from the Hanoi, Phuc Yen areas. COMINT ELINT and photography suggest site VN 64 attempted a missile launch. The initial Fan Song low-PRF detection of 57 nautical miles and the switch into and out of high PRF at 32 and 30 nautical miles were compatible with known surface-to-air missile operations.

Although the surface-to-air missile battalion at VN 64 tracked the mission vehicle in the high-PRF mode for approximately 60 seconds, it did not launch a missile. The offset range between the mission flight path and SAM site VN 64 was 24.7 nautical miles. Hence, a missile launch attempted by this unit posed no possible threat to Collins' A-12.

Chinese and North Vietnamese air defense echelons reported tracking Collins' A-12 reconnaissance plane and apparently gave position plots of the high-flying, fast-moving A-12 a special "trinome," a French word used in algebra to express any quantity produced by the addition of three numbers or quantities that were incommensurable. The Chinese designated the trinome for priority passage of special trinome messages that enabled the Chinese to reduce delay times by 4 to 6 minutes for position plot routing. The elapsed time from initial detection by a Chinese facility to rebroadcast by Hanoi became about 1 to 3 minutes for the Vietnamese SA 2 system to prepare for engagement.

There was no indication of tracking by North Vietnamese air surveillance facilities on this mission. Two Chinese radar stations. Hsi-chi and Chiang-cheng were noted reporting tracking on the vehicle. The Hsi-chi radar station initially detected the vehicle at a ground range of 138 nautical miles and tracked it to 140 nautical miles. Chiang-cheng initially tracked the vehicle at a ground range of 128 nautical miles and carried it out to 254 nautical miles. Seven DOD strike reconnaissance aircraft were active during the surface-to-air missile activity of Collins' first pass, On Collins' uneventful second pass, out of 57 strike aircraft, four ECM Iron Hand aircraft were active.

BX6727

On 6 October 1967, Frank Murray, Dutch 20 flew A-12 Article 131 on BLACK SHIELD mission BX6727, a single-pass reconnaissance mission flown over North Vietnam.

Murray's photography gave no evidence of surface-to-surface missile facilities or equipment. Overall, he recovered good photography despite scattered clouds covering approximately 40 percent of the target area. His single-pass mission photographed thirty-six COMIREX targets, nine of them SA-2 sites with undetermined occupancy.

Murray's System 6 tapes recorded seven Fan Song signals during the mission overflight. None of these signals appeared to be tracking the vehicle, and no onboard defensive systems activated because of these signals.

At 0128 hours ZULU, the surface-to-air missile battalion at VN 142 received an alert to an approaching SR-71. The warning message, "Be ready to do battle," arrived 51 minutes before the initial detection of Murray's flight by Chinese radar facilities. (The CIA acted to determine the ·original source of this forewarning, but results were negative. Such a forewarning was significant in that it provided the North Vietnamese with ample time to ·prepare whatever defenses they might choose to use against the BLACK SHIELD vehicle.) COMINT noted four SAM battalions reacting to Murray's A-12 aircraft. Murray saw no indication of launch activity by any of them. The mission flight path went well north of these surface-to-air missile sites. Two of the battalions located by COMINT at sites 142 and 148, were offset 60 and 65 nautical miles from Murray's track.

Murray noted both the Chinese and North Vietnamese air defense facilities tracking him starting from 0219 hours to 0253 hours ZULU. Bai Thuong was the first North Vietnamese radar station that had it reporting a track of Murray's plane intercepted. Bai Thuong detected the vehicle at 0237 hours ZULU at a ground range of 156 nautical miles and continued tracking it out to a range of 172 nautical miles. Three Chinese radar stations -- San-pao-chi, Chiang-cheng in China, and Dien Bien Phu in North Vietnam-- reported initial detection, and final tracking ranges respectively of 56-248 nautical miles, 142-29 nautical miles, and 180-64 nautical miles. The reported tracks and altitudes were in general compatible with those of Murray's aircraft. During the overflight period, one EB-66C and one EB-66B jamming platforms were active. However, strike activity remained light.

It was after Murray took his photo run that his flight became the CIA's worst nightmare. Murray was flying the previously canceled Mission BX 6726. As he completed his first pass over denied territory, he heard a loud bumping sound and noticed a fluctuation in the oil pressure gauge for his right engine. The oil pressure of the right engine had dropped to zero at he approached 20 miles to INS position 06.

Murray shut down the right engine and kept the good engine at max thrust until clear of North Vietnam airspace. He then started his descent on the path to the intended tanker. While over Laos, Murray called the tanker on the secure radio, telling the tanker crew that he was aborting the rest of the mission and planned to land at Takhli AB in Thailand.

Unknown to Murray, Takhli AB was the home of a CIA U-2 detachment. The tanker relayed what he knew back home to Kadena. However, the abort was well underway by now. When Murray arrived overhead at Takhli, he tried to contact the control tower without success. It was then that the tanker reminded him he was still transmitting in the secure radio mode. This explained why the tower was not answering.

After switching to the standard UHF mode, he told the tower that an emergency recovery was going to happen soon. The tower wanted to know what type airplane was involved, how many souls on-board, and how much fuel on-board. The usual things a control tower asks. Murray told the tower the airplane type was classified and that he had one soul on-board and the fuel state was about 15,000 lbs. About this time the tower was asking if the emergency aircraft could hold while the base launched a large strike force.

Murray told the tower that he was twenty miles out on a straight-in approach. All the while the base was not visible to Murray due to cloud cover over the base. At about five miles and under the cloud deck Murray sighted the runway and prepared for a single-engine landing (which for him was not a great problem).

With the gear down, the tower could now see what must have been quite a sight, an A-12 approaching the base. The landing was uneventful, but as Murray cleared the runway on a taxiway, he informed the tower that the small chute that deployed the main drag-chute had fouled the runway with rubber bands. This unscheduled emergency landing caused a considerable delay in the launch of a large F-105 strike force bound for North Vietnam. Murray's tanker also landed as he had the PF-1 fuel needed to service the A-12 for the recovery to Kadena.

After stopping on the taxiway, Murray called for the base commander to discuss what to do and where to put the A-12. The first officer to come out to the A-12 was the Thai base commander. With that, Murray asked the tower for the American commander, who did arrive shortly after and they discussed where to move the A-12. The good engine was still running, so the airplane could move under its power. They decided that the best place now was the compass rose parking stand off the main taxiway.

It was there that Murray gave the American officer his letter of introduction. Shortly after that a group of people approached the A-12 and identified themselves as CIA people from the U-2 detachment. Now having knowledgeable assistance, Murray could shut down the engine providing air conditioning to his flight suit. The U-2 people indicated that they would try to move the A-12 to their hangar as there were no other hangars on the base, only nose docks. They transported Murray to the U-2 compound to remove the pressure suit and make a call to Kadena describing the airplane problems.

Between the U-2 people and the base, they concocted a tow bar to use in moving the A-12 to the U-2 hangar. The pathway to the hangar was for the U-2, so it consisted of a single narrow hard surface which the A-12 nose wheel rode. The main gear rode on the dirt off the hard surface.

They planned to position PSP planking for the main gear to ride on. Murray was in the cockpit while the towing operation took place. As the A-12 wheels rolled over the PSP, the weight of the airplane destroyed it, but they had plenty of PSP, so they continued to the hangar.

The U-2 hangar was sufficiently wide to house the A-12, but only marginally suited in height. As the A-12 was approaching the hangar, it became clear that the rudders were going to strike the top of the hangar doorway. After much discussion as for how to proceed, they decided to deflate the main gear struts to gain clearance for the rudders. This action worked, and the A-12 was now securely in the hangar with the doors shut to prevent prying eyes from watching what went on. The U-2 detachment people could download the camera and remove the film (take) for processing. Murray borrowed a flying suit from the U-2 crew and awaited the recovery team.

At the first indication of an abort, extending from Kadena Air Base to CIA Headquarters at Langley, a mad scramble had erupted to get Murray to safety and to protect the integrity of the mission. Protecting the secrecy of the A-12 was a prime concern. The compromising of the call signs by an unknown station required the assignment of new call signs for an A-12 "buddy" recovery by Detachment 4 commo. A rescue team of 9 officers, eight enlisted, 33 civilians, and 4000 pounds of equipment were loaded on a KC-135 to join up with Murray and the plane. A C-130 with 24,800 pounds of equipment and two civilians followed. The trip took 4 hours 11 minutes. The problem turned out to be a faulty oil pressure sensor.

Unable to repair the broken link to oil pressure sensor, they ran a wire through a panel door to provide operational oil pressure information. The improvised repair restricted the flight to low speed, which created all sorts of problems as a low and slow flight home to Kadena would expose the Oxcart vehicle to unauthorized viewing and to rapidly deteriorating weather conditions.

After the completing the repairs in the U-2 hangar, they moved the A-12 outdoors back to the compass rose parking stand. The weak taxiway became a problem as it was not strong enough for the plane to return to the hangar after loading with fuel. In the event of a delayed takeoff, they planned to cover the plane with tarps after servicing.

The ground crew did an engine start to check on their work. All went well enough to plan for the flight to Kadena. Part of the preparation was the alignment and loading of the INS route to the home base. The INS crew did not have a good benchmark position to load (present position), so they used the Takhli AB base-ops coordinates furnished by the base.

Normally the A-12 would have been ground run on a tie down before releasing for flight, but there was no tie down good enough for the A-12. Murray tested the engine on takeoff.

Two SAC 903rd KC-135s took up station to organize a return buddy flight. It complicated matters to learn the recovery equipment brought up in C-130 was too large to fit in a KC-135 for a return to Kadena.

The departure of the buddy flight required coordination with the area combat support commander to schedule departure during the period of least observance and between takeoff of combat craft and their return. The return flight had to avoid high traffic density and minimize the opportunity for other aircraft to observe the Oxcart vehicle in flight. Heavy cirrus, ice and CB activity on both return routes became a factor. The Article would travel within a cell with a KC-135 carrying the PF-1 fuel. Article 131 carried enough to make an emergency landing in South Vietnam if in dire need. Article 131 took up the identity of a KC-135 on buddy flight. F-105's from Takhli escorted the cell to a point near Saigon where F-4s took up the escort mission to a safe point off-shore South Vietnam.

There was no need for a pressure suit on the low altitude return flight, so Murray used a summer flying suit and his regular flying helmet that the recovery crew had brought from Kadena. One can only imagine the people gawking as Murray, his KC-135, and F-105 escorts took the runway for departure. Everybody on the base knew there was a Blackbird there and when the commotion started with the start cart Buicks running, they lined the taxiways. After liftoff of the buddy flight, the CIA security people went crazy trying to round up the cameras on base.

As the weather worsened, the CIA decided to make the KC-135 the buddy flight leader so it could use its radar to vector the flight around the weather cells.

Sometimes on the flight back, the tanker would lead, and other times Murray led the trip. They had to not only avoid the storms but a B-52 Arc Lite flight in Philippine airspace as well.

The return required three aerial refueling of Murray's A-12. The return trip flew at 30,000 feet. The INS worked well and never updated. When Murray landed back at Kadena, the INS was within spec for position and ground speed. Murray had flown at Mach 3.19 and 81,000 feet for a duration of 2:20 hours before aborting. Imagery quality was good. His buddy flight back to Kadena lasted over 5 hours. The recovery and the experiences learned became procedure and part of the TAC doctrine. The CIA's wandering boy returned with a sore butt and credit for half a mission.

Stemming from this event was a couple of interesting side stories of things that happened along the way. The first was something that occurred after a couple of days of Murray lounging around the U-2 compound when he asked to borrow some civilian clothing to go to the Officer's Club on the Base. The CIA U-2 detachment people were not fond of that idea but said OK so long as one of their security people went along. So off they went to get a beer. They were sitting at the bar when Murray felt a tap on the shoulder. Turning to see who did the tapping, Murray was surprised to see an old friend from the 60th Fighter Interceptor Squadron; the unit Murray came from to go to Area 51 and the Oxcart Program. The old friend asked what Murray was doing at Takhli, so Murray told him he was a Hughes Tech Rep, working on radar in the vicinity. The friend exchanged chit chat and went his way.

Sometime later, after Murray went back in the Air Force, he was sent to a refresher course on Sea Survival at Tyndall AFB in Florida. It was at this school that the old friend found Murray once again, this time as a lieutenant colonel in the air force. Murray told him that he just decided to get back in the Blue Suit. The old friend said something about doubting anything Murray said, quipping that he must have been the guy that flew the Blackbird into Takhli back in '67. Murray could not confirm anything since the Program was still classified. Now one can see why the CIA guys did not want Murray to go to the Club at Takhli.

The second interesting event was the effect of having seen Murray and his A-12 in a combat zone. The security guys had the Vice Wing Commander of the F-105 fighter squadron sign a paper where he was required to uphold the security of the Oxcart operation. This effectively ended the colonel's combat flying in that war. Ironically, the wing commander Colonel was one John Giraudo. His brother Joe had been at Area 51 for some time working for the CIA as a blue suiter on Project OXCART.

This abort mission was one of a kind during the BLACK SHIELD operation; all others recovered to Kadena after overflights.

BX6728

On 15 October 1967, Kenneth Collins, Dutch 21, flew A-12 Article 131 on BLACK SHIELD mission BX6728 on a double-pass, high-altitude reconnaissance mission flown over North Vietnam.

Extensive clouds limited clear coverage to about 10 percent of the photographed area. Consequently, mission photography provided no evidence of surface-to-surface missile equipment or facilities. Nonetheless, Collins photographed 19 COMIREX targets and one new site. He found none of the seven SA-2 sites occupied.

From 0025 hours ZULU to 0111 hours ZULU, authoritative Chinese echelons passed warning messages to Chinese radar stations. These messages informed the stations, "Today there was a high-altitude, high-speed situation. Complete preparations." Thus, Collins' BX6728 mission became the second BLACK SHIELD mission to be preceded by a forewarning. The correlation of the time of intercept of the BX6728 warning message with that of BX6727 indicated a probable tip-off from the South China Sea refueling area. The CIA employed a more secure communications posture in the refueling area following Collins' flight and never noted another forewarning message.

The System 6 recorded one Fan Song signal during the overflight period that did not appear to be emanating from a site tracking Collins' A-12 and his missile defensive mechanism never activated because of this intercept.

Collins noted the Chinese and North Vietnamese air defense facilities reflecting tracking data on him from 0323 hours ZULU to $033 hours ZULU. The latter monitored the Kun-ming sector broadcast and rebroadcast his track over the Hanoi Bae Mai facility. The available information indicated little or no strike jamming operations during the overflight period.

BX6729

On 18 October 1967, Frank Murray, Dutch 20, flew A-12 Article 129 on BLACK SHIELD Mission BX6729, a double-pass reconnaissance mission flown over North Vietnam. A display of the vehicle flight route and air surveillance tracking surface missile search area, but there was no evidence of surface-to-surface deployment.

Murray's photography noted construction activity by Chinese engineers of Yep. Bai Airfield. He photographed 193 COMIREX targets, 13 of them SA-2 sites, and 16 of them occupied including one newly identified site. Dummy SA-2 missiles occupied one SA2 site. The System 6 recorded one Fan Song signal. However, the signal was not tracking the aircraft. Therefore, no onboard defensive systems activated during the overflight.

Murray noted Chinese and North Vietnamese air defense facilities reporting tracking of the A-12 with inclusive plot times from 0334 hours to 040 hours ZULU, and from 0503 hours to 0533 hours ZULU. The Chinese radar station Chiang-cheng detected and tracked Murray's plane at radar station vehicle to ground ranges of 100-240 nautical miles and 124-253 nautical miles respectively. The North Vietnamese radar station at Vinh tracked and rebroadcasted the track of the A-12. Murray failed to note any strike jamming aircraft in operation during the first pass, but he did note two EB-66 active jamming platforms in operation during the second pass. Overall signal density was light.

BX6732

On 28 October 1967, Dennis Sullivan, Dutch 23 flew A-12 Article 131 on BLACK SHIELD mission BX6732, a double-pass, high-altitude reconnaissance mission flown over North Vietnam. The enemy fired a missile at Sullivan's A-12 during the second pass.

Sullivan's photography gave no evidence of surface-to-surface missile deployment. Mainly, he saw where the probable rail-to-road transshipment point appeared under construction near Ho-kou, China, across the border from Lao Cai, North Vietnam. His photography provided good coverage of all six major airfields in North Vietnam that all except for Haiphong Cat B Airfield appeared serviceable. He photographed 167 COMIREX targets and seven bonus targets, 120 North Vietnamese SA-2 sites and 14 of them occupied. During the first pass of the overflight, his System 6 recorded four Fan Song signals, none of them emanating from radars tracking his A-12. The signals failed to activate his onboard defensive systems.

During the second pass, his System 6 recorded three Fan Song signals, two of them tracking him. Correlation of all information of the overflight showed where the North Vietnamese surface-to-air missile site VN 133 launched a single, albeit unsuccessful, missile at his plane. This was the first known missile firing at a BLACK SHIELD vehicle by the North Vietnamese surface-to-air missile air defense system. COMINT indicated that site VN 133 and an unidentified site were tracking Sullivan and that the former launched one missile.

Analysis of COMINT and System 6 tapes gave evidence of a missile launch while the Fan Song radar was in the low-PRF mode (1,200-1,265 pps – pulses per second). Before this mission, all known launches by S-2 systems occurred when the Fan Song radar was in the high PRF (pulse repetition frequency) mode (2,400-2,530 pps). The low-PRF launch mode required some modification in the SA-2 system most likely performed in-the-field. This change in procedure attempted to cope with a Mach-3 target. The SA-2's system design was for a maximum target velocity of Mach 2. Two photographs from this mission showed missile smoke above site VN 133 and the other of a missile (with missile vapor trail). The photograph also shows the missile flight path was then down and away from Sullivan's BLACK SHIELD vehicle flight path.

The low PRF launch mode extended the launch window of the SA-2 system. However, even though this tactic maximized system capability against a non-jamming Mach-3 target, the kill probability with the ECM equipment now on-board the BLACK SHIELD vehicle remained essentially unchanged. The ECM equipment appeared to perform well against this first firing at the mission aircraft. While the low-PRF launch mode allowed missile firings at ranges greater than the normal high-PRF launch range of 32 nautical miles, this missile launch occurred at a range of 25.2 nautical miles. The regimental controller criticized the surface-to-air missile battalion controller for launching late, especially after the regimental controller ordered a firing at 32.5 nautical miles.

Both the by Chinese and North Vietnamese air defense facilities carried out the air surveillance tracking of Sullivan's A-12. The tracking was, in general, complete, and accurate. DOD strike jamming reports indicated that 23 aircraft were conducting operations during the first pass overflight. During the missile launch of the second pass, three DOD strike aircraft were conducting operations.

BX6733

On 29 October 1967, Frank Murray, Dutch 20, flew A-12 Article 127 on BLACK SHIELD mission BX6733, a two-pass reconnaissance mission flown over North Vietnam.

Murray's photography gave no evidence of surface-to-surface equipment facilities. However, he saw continuing construction by Chinese engineers of North Vietnamese Yen Bai Airfield. This was the first sign of progress since 20 June 1967. Murray photographed 120 C0MlREX targets, 74 of them SA 2 sites and 16 of them occupied. He identified the one new unoccupied SA-2 site. His System 6 recorded eight Fan Song signals while his aircraft was over North Vietnam. However, none of the signals appeared to be tracking him. His onboard defensive systems never activated because of these signals.

During his first pass, Murray saw no indication of air surveillance tracking by either the North Vietnamese or the Chinese. Nor did he report any strike jamming operations.

The Chinese air surveillance facilities tracked the BLACK SHIELD vehicle continuously during the second pass. However, there was no evidence of air surveillance tracking by the North Vietnamese. However, the Chinese did pass the BLACK SHIELD track to the North Vietnamese. Moderately heavy strike jamming operations by 38 aircraft occurred during this pass. Both Chinese radars stations at Chiang cheng and Ho-lo-shih passed mission plots. The initial detection and final tracking ranges for the Chiang cheng radar station were 80 and 228 nautical miles respectively. The Ho-lo-shih radar station's detection and final tracking ranges were 175 nautical miles and 265 nautical miles. Correlation of System 6 with the final position plot of the Holo-shih radar station indicated the tracking radar to be a Moon Face VHF radar. The Phuc Yen regimental controller and subordinate surface-to-air missile site VN 234 were ·noted in ·communications referring lo Murray's BLACK·SHIELD vehicle, which was over the Gulf of Tonkin at the time it was never in VN 234's firing zone.

BX6734

On 30 October 1967, Dennis Sullivan, Dutch 23, flew A-12 Article 129 on BLACK SHIELD Mission BX6734, a high-altitude, double--pass reconnaissance mission flown over North Vietnam.

Sullivan photographed 118 COMIREX targets, 92 of them North Vietnamese SA-2 sites, 13 of them occupied and one a newly identified site. Sullivan's flight saw no evidence of surface-to-surface facilities or deployment. ·

He noted surface-to-air missile reactions on the first pass.

During the first pass, two surface-to-air missile battalions located at sites VN 234 ad VN 12 reflected an attempted missile launch. Neither of these sites launched any missiles, both being out of the required range limits for a successful launch. Analysis of Sullivan's System 6 tapes indicated a low-PRF launch, like the one during mission BX6732. Signal analysis coupled with the missile fly out characteristics indicated the possible launching site to have been VN 167.

Sullivan did not report sighting a vapor trail or missile, and it was possible the site fired at another aircraft. The intercepted Fan Song signal was weak. Thus, the Mad Moth jammer stayed active for only a short period,

Correlation and analysis of all available data on the second pass showed that at least six surface-to-air missile sites fired at Sullivan's A-12 from eight to ten missiles in a 2-minute period. As might be expected the dense signal environment during the multiple engagements made a complete delineation of events difficult. Six active Blue Dog channels which tend to indicate a minimum of six missile launches most probably carried out by these six surface-to-air missile sites. BLACK SHIELD photography ·shows vapor trails of six missiles. Sullivan also saw these vapor trails and witnessed three SA-2 detonations.

The missile launches were at ranges of 36, 38, and 41 nautical miles. Since these launch ranges were beyond the high-PRF acquisition range of the Fan Song radar (32.6 nautical miles), the launches might be construed to have been low-PRF launches. The ECM equipment (Mad Moth and Blue Dog) appeared to function normally. Mad Moth and Blue Dog replied to the signal environment and type of interference on S-band were experienced by operators at three of the SA-2 sites. Sullivan did not report any degree of jamming evident, but the fact that the aircraft did not sustain a hit by a warhead pellet in an eight-missile-launch environment appeared to be a measure of the ECM configuration's effectiveness.

Post-flight inspection of Sullivan's A-12 revealed that a piece of metal penetrated the lower right wing fillet area and lodged against the support structure of the wing tank. The fragment was not a SA-2 warhead pellet. It was apparently of Soviet manufacture. It was possibly a part of the debris from one of the three SA-2 missile detonations observed by the pilot. Spectrographic and other measurements made on the fragment and various components of a SA-2 MOD I missile to date have shown no correlation. The shrapnel is now on display at the CIA Museum at Langley.

Chinese and North Vietnamese air defense facilities continuously tracked the mission vehicle over North Vietnam and the Gulf of Tonkin.

Sullivan reported the DOD strike jamming activity as light during the first pass. In contrast. 62 aircraft conducted moderately heavy strike jamming operations in the Haiphong area during Sullivan's second pass. However, the surface-to-air missile launches happened near Hanoi.

BX6737

On 8 December 1967, Mele Vojvodich, Dutch 30, flew A-12 Article 131 on BLACK SHIELD mission BX6737, a high-altitude, single pass reconnaissance mission flown over the Cambodian South Vietnamese and Laotian North Vietnamese border areas.

Vojvodich's mission looking for evidence of Viet Cong or Vietnamese activity covered seven specified priority search areas in northeastern Cambodia, southeastern Laos, and adjacent South Vietnam.

Vojvodich's photography provided evidence of expanded supply transshipment facilities along the Tonle Kong River in Cambodia near the Laotian border. He detected several troop encampments served by numerous trails extending north to Cambodian Route 97 and identified a probable new storage area and water to road transshipment point. His System 6 recorded no threat signals, and he saw no active defensive systems during the overflight. Chinese radar facilities conducted ten minutes of air surveillance tracking. However, Vojvodich saw no indication of North Vietnamese awareness of the mission.

BX6738

On 10 December 1967, Jack Layton, Dutch 27, flew A-12 Article 131 on BLACK SHIELD mission BX6738, a single pass reconnaissance mission flown over the Cambodian, Laotian, and South Vietnamese tri-border area.

Layton's Mission photography detected a new probable transshipment storage facility on the Tonle Son River in Cambodia near the South Vietnamese border. Photography also showed a heavily used trail extending from Cambodia into the northwestern corner of Darlac Province in South Vietnam. He observed two automatic weapons positions near the trail just inside the Cambodian border. He recorded no threat signals or saw any defensive system activated during the overflight. He saw no evidence of air surveillance tracking by either the Chinese or North Vietnamese.

BX6739

On 15 December 1967, Mele Vojvodich, Dutch 30, flew A-12 Article 127 on BLACK SHIELD Mission BX6739, a double pass, high-altitude mission flown over North Vietnam.

Vojvodich obtained clear and interpretable mission photography from this two-pass, cloud-free mission. He saw no evidence of surface to surface missile deployment. All North Vietnam's major airfields, except Haiphong Cat Bi, appeared serviceable.

Vojvodich photographed 195 COMIREX targets, 142 of them North Vietnamese SA 2 sites and 18 of them occupied, His System 6 recorded three Fan Song signals during the overflight with none of them tracking the A-12, None of his defensive mechanism activated because of the signals. Both the Chinese and North Vietnamese air defense facilities tracked the A-12. BaThuong, a North Vietnamese radar station reported two position plots of the A-12 at the radar to aircraft ground ranges at 180 and 148 nautical miles, respectively. DOD strike jamming operations occurred during the overflight.

BX6740

On 18 December 1967, Jack Layton, Dutch 27, flew A-12 Article 131 on BLACK SHIELD Mission BX6740, a double pass reconnaissance mission flown over North Vietnam on 16 December 1967

Mission photographs obtained about 60 percent cloud-free coverage of the target area. Layton saw no evidence of surface-to-surface missile deployment.

The mission aircraft photographed 86 COMIREX targets, 74 of them SÁ-2 sites and two then occupied. Missions BX6739 and BX6740, flown on consecutive days, photographed 221 of North Vietnam's 226 useable SA 2 sites, including six new locations. Twenty were occupied, including five of the six new sites. The System 6 recorded six Fan Song signals, none of which appeared to be tracking the vehicle. No onboard defensive system activated. Chinese air surveillance facilities tracked the mission vehicle, and a Chinese radar station at Pa ka reflected tracking the aircraft at the radar to aircraft ground ranges of 134 and 234 nautical miles, respectively. There was no indication of North Vietnamese awareness of the mission vehicle. Moderate DOD strike jamming operations occurred during the overflight.

BX6842

On 4 January 1968, Jack Layton, Dutch 27, flew A-12 Article 127 on BLACK SHIELD mission BX6842, a single-pass, high-altitude reconnaissance missions flown over North Vietnam where he encountered a SAM reaction. He saw no indication of surface-to-surface missile (SSM) activity in the photographed areas. He did, however, photograph 176 Chinese and North Vietnamese COMIREX targets plus 16 bonus (non-COMIREX) targets. Of the 176 COMIREX targets photographed, 139 were North Vietnamese SA-2 sites, and 12 of them occupied. The mission vehicle also photographed several probable cave defense sites near Hanoi and Haiphong.

Layton recorded three S-band and one L-band Fan Song B radar signals during the overflight. The third S-band signal qualified as a valid threat and activated the Mad Moth and Blue Dog BLACK SHIELD jammers. Correlation and analysis of all available data indicated that SAM site launched a missile at Layton's A-12. Of three photographs, one showed smoke over site 267; the second of a launched missile; a missile vapor trail and missile burn-out; and the third a missile detonation. Missile trajectory analysis from photography indicated the missile's approach to the mission vehicle was no closer than 8,000 feet and probably was much greater.

This mission noted a departure ·from the normal Fan Song B launch guidance technique. Normal operational procedures specified launch guidance to occur while the target and missile were under the influence of the Fan Song B radar operating in the high-PRF mode. BLACK SHIELD missions began observing intentional atypical missile launch sequences in October 1967 during missions BX6732 and BX6734 with the launching of missiles while the radar was in low PRF. The Fan Song switched to high PRF in two of these firings approximately six seconds after launch. It applied the missile guidance after that. The missile fired at the Layton's A-12 during BX684 launched with the radar in the low-PRF ·mode, and missile guidance (high-PRF) information computed with the radar operating in the low-PRF mode. This atypical launch/guidance sequence might have been an attempt by the Vietnamese to reduce the effect ·of the BLACK SHIELD jammers while suffering only a small loss in capacity due to the lower data rate.

The situation geometry indicated that the launch occurred in anticipation of a· BLACK SHIELD path offset amendable to a successful missile intercept. Erroneous EW tracking information generated this prediction which· projected the target flight path to within 4 nautical miles of the SA-.2 site. The actual vehicle flight path was approximately 20 nautical miles from site 267 and too distant for a successful intercept. The erroneous EW tracking data coupled with the effects of the BLACK SHIELD jammers resulted in a late and probably hasty launch, a launch that could not have been successful because of the large offset range.· The missile continued to automatic destruct which occurs approximately 62 seconds after launch. This was also the approximate duration of the L-band signal.

BX6843

On 5 January 1968, Jack Weeks, Dutch 29, flew A-12 Article 131 on BLACK SHIELD Mission BS6843, a double-pass overflight of North Vietnam.

Week's mission called for photographing two suspect SS (cruise) missile sites near Thanh Hoa, one containing a few small unidentified objects and the other unoccupied. Weeks photographed 233 Chinese and North Vietnamese •COMIREX targets plus 10 bonus (non-COMIREX) targets. Of the 23 COMIREX targets photographed, 182 were North Vietnamese SA-2 sites with 15 of the sites occupied. Week's mission also provided photography of five of North Vietnam's major airfields and coverage of almost all the rail network. Weeks recorded eleven Fan Song B S-band signals during the overflight with none of them appeared to be tracking the A-121. His mission activated no defensive systems even though the Chinese accomplished initial EW tracking and the North Vietnamese broadcasted the track information. Weeks saw no Indication of tracking by North Vietnamese radars. Initial detection and final track ranges of Chinese radar stations during the first pass were 109 and 217 nautical miles for Chiang-cheng and 96 and 237 nautical miles for Pu-kao. Chiang-cheng's ranges for the second pass track were 117 and 325 nautical miles. There was little or no strike jamming during the overflight.

BX6847

On 26 January 1968, Jack Weeks, Dutch 29, flew A-12 Article 131 on BLACK SHIELD mission BX6847, a three-pass, high-altitude reconnaissance mission flown over North Korea on 26 January 1968.

Week's A-12 photographed 82 North Korean and Chinese COMIREX targets and 837 bonus targets. His mission obtained comprehensive baseline coverage of most of North Korea's armed forces and industry as well as large portions of the transportation system. Week's photography observed the USS Pueblo and three guided missile patrol boats (Komar PTG) on this mission. He obtained a photograph of the USS Pueblo at anchor in a bay north of Wonsan.

Week's mission saw no indication of a hostile weapons reaction and none of the A-12's defensive systems activated during the overflight. Chinese radar facilities at Mu-yeh Island, Hai-yang Island, and Hunch'un did most the air surveillance tracking. Analysis of System 6 ELINT indicated that Soviet Bar Look, TALLINN, and Side Net radars, subordinate to Uglovoe, tracked the vehicle. The Soviets broadcast one position plot of the vehicle, but there was no indication of any North Korean surveillance tracking.

During this mission, Weeks was experiencing engine problems that made it difficult to maintain the mission altitude of 80,000 feet. He returned from his flight in an abort condition because of overheating the A-12's J58 engines while attempting to maintain altitude. Known only at CIA and NRO-level, the Chinese tracked Week's flight and so informed North Korea. Also, known at that level was the failure of the first flight to cover certain areas of interest to Washington concerning possible hostile moves by China.

BX6851

On 16 February 1968, Kenneth Collins, Dutch 21, flew A-12 Article 127 on BLACK SHIELD mission BX6851, a two-pass mission flown over North Vietnam. A rapid deterioration of weather over the target area resulted in little usable photography.

Collins saw no indication of a hostile weapons reaction, and no electronic defensive equipment activated during the overflight. Chinese radar at Pei-li and Chiang-cheng maintained air surveillance tracking of his mission. Initial detection and final radar-to-aircraft track range for these radar stations were 39 to 228, 126 to 286, and 169 to 233 nautical miles, respectively, although there was no indication of North Vietnamese radar tracking, track information probably was obtained by monitoring Chinese broadcast facilities. There was little or no strike jamming activity during the overflight period.

BX6853

Following the Jack Weeks' BX6847 mission over North Korea on 26 January 1968, known only at CIA and NRO-level, the Chinese had tracked the first flight and so informed North Korea. Also, known at that level was the failure of the first flight to cover certain areas of interest to Washington concerning possible hostile moves by China. Department of State was reluctant to endorse a second mission over hostile territory for fear of diplomatic repercussions. However, NRO insisted it needed a second flight. Thus, BGen Paul Bacalis briefed Secretary Rusk who decided to alter the flight plan to obtain the data needed by NRO. On 19 February 1968, Frank Murray made this flight, following much of the same flight pattern as Jack Weeks 2 weeks earlier. As with Jack Weeks, only Murray and the CIA Flight Planner knew the pattern, mission, and results at the operational level.

Frank Murray, Dutch 20, flight in A-12 Article 127 was, a two-pass reconnaissance mission flown over North Korea.

Murray photographed 84 North Korean COMIREX targets plus 89 bonus targets. Scattered clouds covered 20 percent of the area that Jack Weeks' mission BX6847 photographed, concealing the location of the USS Pueblo. Murray identified one new occupied SA-2 site near Wonsan. However, he saw no indication of a hostile weapons reaction through the Chinese air defense system (ADS) air surveillance did track the A-12. Murray's mission marked the first tracking of the A-12 by the North Korean ADS. The North Korean Maryong-San radar station reported-track and altitude information on his second pass. Initial detection and final track ranges for the Maryong-San radar station were 54 and 237 nautical miles. Murray saw no indication that any Soviet radar systems tracked his mission.

BX6856

On 8 March 1968, Mele Vojvodich, Dutch 30, flew Article127 on BLACK SHIELD Mission BX6856, a two-pass mission flown over North Vietnam and the Demilitarized Zone.

Vojvodich obtained good quality photography of the Khe Sanh and the Laos, Cambodia, and South Vietnamese border areas. No useable photography resulted from the North Vietnam mission due to adverse weather conditions. There was no indication of a hostile weapons reaction, and no onboard defensive systems were activated. Chinese Air Defense radar facilities tracked Vojvodich's A-12. However, there was no apparent tracking of the vehicle or indication of knowledge of the overflight by the North Vietnamese. A change in normal route penetration apparently hampered detection of the vehicle by Chinese/North Vietnamese radar facilities. Little or no strike/jamming activity occurred during the overflight.

BX6858

On 5 May 1968, Jack Layton, Dutch 27, flew A-12 Article 127 on BLACK SHIELD Mission BX6858, a double-pass, high-altitude reconnaissance mission flown over North Korea. Cloud cover and heavy haze severely hampered interpretability of Layton's mission photography. Nonetheless, the mission aircraft photographed 68 COMIREX targets plus 30 bonus targets. Of the targets photographed, 15 were SA-2 sites with three of them occupied, one unoccupied, and eleven identified only. The mission tentatively identified a possible SAMLET coastal defense cruise missile site on the east coast between Wonsan and Ham hung. Existing weather conditions in the target area did not permit the photographing of the USS Pueblo.

Layton saw no indication of a hostile weapons reaction. Elements of the Chinese Air Defense System (ADS) accomplished air surveillance tracking of Layton's A-12 on both passes. Initial detection and final track ground ranges for the Chinese radar station at Mata Point were 118 and 264 nautical miles, respectively. There was no indication of tracking by either the North Korean ADS or the Soviet ADS.

The number of radar signals received by Layton's BLACK SHIELD Mission BX6858 was:

Big Mesh	2	Cross Slot	2
Side Net	2	Bar Lock	11
Rock Cake	4	Moon Cone	9

Cross Legs	1	Moon Face	3
Token	3	Long Talk	1
One Eye	1		

Chapter 12 - Mission Analysis.

The six CIA projects flew 29 missions, 24 over North Vietnam, two over Cambodia, and three over North Korea, accomplishing fifty-eight photographic flight, line. All missions launched and recovered from Kadena, except for one where Frank Murray made a precautionary landing at Ban Takhli, Thailand. All but four missions reported enemy radar tracking that ranged from very brief reflections of the A-12's presence on early missions to extended and accurate tracking.

A Missile Strike.

On three missions, the enemy unsuccessfully launched SA 2 missiles at the A-12. Post-flight inspection of the mission aircraft of Dennis Sullivan's 30 October 1967 revealed that a piece of metal had penetrated the lower wing surface of the aircraft. It was not an SA 2 warhead pellet, but possibly a part of the debris, from one of the missile detonations. Sullivan reported eight missiles launched during the mission,

The Arrival of the Air Force's SR-71.

The initial SR-71 cadre was formed at Beale AFB in January 1965 and was designated the 4200th Strategic Reconnaissance Wing. A year later, Lockheed's Advanced Development Projects (Kelly Johnson's Skunk Works) delivered the first of the high technology SR-71 Strategic Reconnaissance Wing aircraft.

The SR-71 was declared operationally ready in 1968.

In early March 1968, the air force's SR-71 aircraft arrived at Kadena to relieve the OXCART Detachment of its BLACK SHIELD commitment for North Vietnam coverage.

The OXCART Detachment went on standby status to back up the SR 71 capability.

On a humorous note, when the SR-71 Blackbirds arrived at Kadena, the personnel had no idea of the A-12 even existing. They saw American civilians living in Morgan Manor and the A-12s in the CIA hangars. They watched them take off and return as the pilots made regular pilot currency flights, however, the CIA refused to allow the air force to enter the area or ask about them. The air force SR-71 personnel referred to the CIA A-12 planes as "Brand X."

In January 1990, the SR-71 program was officially retired. Due to a void in reconnaissance capability, Congressional direction in October 1994 mandated that three SR-71s were to be reactivated and returned to service. The Air Force decided that two SR-71As and NASA's assigned SR-71B would make up the three aircraft cadre. The first of the two refurbished A models by Lockheed-Martin had its initial flight in April 1995 and both were assigned that June to the newly formed Air Force unit, Det 2, 9 SRW, at Edwards AFB, California. In January 1996, the unit was declared operationally ready for worldwide deployment within 24 hours notice. The Detachment flew 150 sorties, participated in several exercises, and flew over 365 hours before the SR-71 program was line-item vetoed by President Clinton in October 1997. During the reactivation, both aircraft were modified with a data link system that permitted near real time transmission to a ground site of its EIP (electro-optical) camera systems. The last SR-71 flight was in October 1999 with NASA crews. NASA officially terminated their program December 2001 and returned the remaining aircraft to the air Force for disposition.

The SR-71, referred to as the Blackbird or the HABU, and its pilots as sled drivers, served with the U.S. Air Force from 1964 to 1998. A total of 32 aircraft were built; 12 were lost in accidents and none lost to enemy action. The SR-71 was designed for flight at over Mach 3 with a flight crew of two in tandem cockpits, with the pilot in the forward cockpit and the Reconnaissance Systems Officer (RSO) operating the surveillance systems and equipment from the rear cockpit, and directing navigation on the mission flight path.

Operational highlights for the entire Blackbird family (YF-12, A-12, and SR-71) as of about 1990 included:

3,551 mission sorties flown
17,300 total sorties flown
11,008 mission flight hours
53,490 total flight hours
2,752 hours Mach 3 time (missions)
11,675 hours Mach 3 time (total)

Cover Stories

When the CIA's A-12s arrived at Kadena for Operation BLACK SHIELD, the CIA needed a cover story to explain their presence. The cover story became the aircraft at Kadena being experimental testbed versions of the YF 12A and SR-71 family, brought there to undergo environmental and field testing. The CIA prohibited the air force SR-71 personnel seeing or asking anything about the Blackbirds that they saw landing and taking off on sorties as the SR-71 arrived. The air force referred to the CIA planes as "Brand X."

It was a month before the first local newspaper first reported the presence of the OXCART vehicles at Kadena, the air force used the CIA's cover story in its news release response to an Okinawan press inquiry. After the initial articles, the Far East newspapers reported nothing significant when referring to the presence of the aircraft on Okinawa. The media and press always used the term. "air force YF 12A/SR-71 planes."

The Kadena detachment prepared to return to Area 51 before the 8 June 1968 date established by project headquarters once it received reaffirmation of the OXCART phase-out decision. After that, the CIA limited any A-12 flights at Kadena to those essential for flying safety. After Jack Layton's flight over North Korea, in May, the CIA was out of Mach 3 reconnaissance business. The CIA pilots now flew just enough to maintain their pilot proficiency for ferrying the three A-12 aircraft from Kadena and to Palmdale, California. Instead of returning home to Area 51, they landed at Palmdale for "sardine-like" storage alongside the Area 51 articles already there with the last leaving Area 51 on 7 June 1968.

Outrunning the Sun

Flying over three times the speed of sound presented some unique opportunities to the Oxcart pilots. Layton recalled flying a mission out of Groom Lake in an A-12, and he flew to a point about 380 miles east of St. Louis, Missouri and made a 180-degree turn and flew back. He had gotten off to quite a late start due to some maintenance delays, so the sun was already low in the sky. On the way out, just before Layton made the turn it started getting dark very rapidly because he was outrunning the sun. Before he completed the turn, it was pitch black. On the way back at about Denver, Colorado the sun popped back up in the sky on the western horizon — opposite to the usual sunrise occurring in the east. Layton arrived back at Groom Lake, in time to watch the sun as it set again. Flying close to 2,200 mph, Layton was outrunning the sun by about 1,200 mph.

Incidents along the way

Repeating Frank Murray's words, "Most of the time, the A-12 operated like a Lady, BUT at times she could be a Bitch." On one mission, as CIA pilot Jack Layton approached the top-off tanker, the Boomer told Layton, "I hope you are not planning on flying fast." Looking down on the A-12, the boom operator could see the A-12 missing a lot of wing skin (chine skins).

Layton aborted the mission and headed to Kadena at lowest practical speed. On his let down to land, the air conditioning system was supercooling, and he had extreme fog in the cockpit. F-102s out of Naha Air Base scrambled and joined up with Jack to aid his recovery. The F-102s provided cues to help Jack get his fogged-up A-12 back to Kadena.

On a functional test flight, Frank Murray experienced a problem that developed on one engine as he approached cruise speed. On the final point in the climb/acceleration, his right engine failed to do the correct shift of the bleed bypass until Mach 3.0. This change should have happened at 2.5 MN. The rest of the cruise out was normal, but on the deceleration on a letdown, the engine failed to come out of bypass, this leading to a compressor stall and engine flameout as the airplane slowed to subsonic.

The engine restarted, and the plane landed normally. Following this incident, the CIA, air force, Lockheed, and most likely Pratt & Whitney decided to change the engine to correct the problem. They set the plane up for a functional test flight two days later.

Two days later, on 4 June 1968, Jack Weeks took Aircraft No. 129 on a check flight made necessary by the change of the engine. The birdwatcher heard from Weeks when 520 miles east of Manila. Then he disappeared. Search and rescue operations found nothing. No one ever ascertained the cause of the accident, and it remains a mystery to this day. Again, the official news release identified the missing aircraft as an SR-71 and security maintained.

The taxi and takeoff proved uneventful, as evidenced by the reception of the required Birdwatcher "Code A" transmission and the lack of any HF transmissions from the pilot. Refueling, 20 minutes after takeoff, was normal. At tanker disconnect, the A-12 was airborne 33 minutes.

The tanker crew observed the A-12 climbing on course in a typical manner in what turned out being the last visual sighting of the aircraft. Kadena received no further communications until 19 minutes later when a Birdwatcher transmission indicated right engine EGT exceeded 860 degrees C. Seven seconds later; Birdwatcher indicated the right engine fuel flow was less than 7500 pounds per hour and repeated that EGT exceeded 860. Eight seconds later, Birdwatcher indicated the A-12 below 68,500 feet and repeated the two previous warnings in its final transmission.

Kadena attempted to contact Weeks by HF-SSB, UHF, and Birdwatcher without success. The operation of the recording and monitoring facilities at the home base continued until the aircraft exhausted its fuel with no further transmissions received. The CIA declared the plane is missing some 500 nautical miles east of the Philippines and 600 nautical miles south of Okinawa. The accident report stated that no recovery of any wreckage of aircraft number 129 (60–6932). Jack Weeks was a religious man with a great family. Sharlene and their children still remember him each year along with his contemporaries on the day of his death over half a century ago.

The loss of Weeks occurred during the recovery of the A-12s back to Area 51 at the shutdown of Operation BLACK SHIELD. Two A-12s remained after the loss of Jack Week's airplane. One of them flown by Dennis Sullivan arrived home to Area 51 without a hitch.

The remaining airplane launched with Ken Collins in the cockpit. He was well on his way when he discovered a fuel leak, forcing him to make an unplanned landing at Wake Island. A recovery crew dispatched to Wake repaired the fuel leak in a few days. Collin's tanker landed with him, so he refueled on the ground, and they took off for a low/slow flight to Hawaii.

Ken Collins remembers heading home with a top-off tanking near Iwo Jima; then the route went by Wake Island, Hawaii and on to home. Unfortunately, 131 had problems along the way. Somewhere between the top-off tanker and Wake Island, a fuel leak developed in the right engine nacelle. The leak was noted by an abnormally high fuel flow and was visible through the canopy periscope, so Ken landed at Wake Island, rather than the planned inflight refueling nearby.

A recovery crew was dispatched to fix the leak and get the Article on to Hickam Air Base in Hawaii, where it could receive full servicing. It took several days for the recovery team to affect the repairs. Ken then flew 131 low and slow with a tanker to top off his tanks from time to time and made it to Hickam safely. Ken left to head for the mainland to attend Jack Weeks memorial service while Frank Murray stayed with 131 servicing for a high and fast return flight.

After the refueling near Hawaii on the first attempt to go home 131 developed the fuel leak from the same part of the engine fuel system. Frank dumped fuel and returned to Hickam. The crew once again replaced the broken fuel manifold, and Frank once again headed home with the usual top-off tanking. On the climb-out from the tanker, the same thing happened, a bad fuel leak from the same side.

This time the recovery crew took extraordinary steps to get the A-12 ready. After repairing the fuel leak, they installed vibration sensors for the ground run and discovered that the airframe mount accessory drive (AMAD) was unbalanced and required change-out. This took several days to do, so Frank lounged around Oahu, enjoying the sights. On the third attempt to leave Hickam, the airplane performed normally so Frank headed home to Area 51. On the way, the Exhaust Gas Temperature gauge for the right engine failed to zero, but that was no reason to abort. Frank matched the engine control for the good engine with those of the left engine and finished off the flight to the Area.

The Second and Final Demise of the CIA's Manned Aerial Surveillance

Despite all the CIA's belated efforts, the secretary of defense on 16 May 1968 reaffirmed the original decision to terminate the OXCART Program and store the aircraft. On 21 May 1968, the president confirmed Secretary Clifford's decision at his weekly luncheon with his principal advisors.

Mr. Helms expressed his reactions to these alternatives in a Memorandum to Messrs. Nitze, Hornig, and Flax, dated 18 April 1968. In it, he questioned why, if storing eight SR-71s was one option, why not store them in all the options, with the resultant savings applied in each case. Helms questioned the lower cost figures of combining the A-12 with the SR-71's and disagreed, for security reasons, with collocating the two fleets. Above all, he felt the desirability of retaining a covert reconnaissance capability under civilian management the key point. He felt such a requirement existed and recommended OXCART continuing at its base at Area 51 under CIA control.

Despite all these belated efforts, the secretary of defense on 16 May 1968 reaffirmed the original decision to terminate the OXCART Program and store the aircraft. On 21 May 1968, the president confirmed Secretary Clifford's decision at his weekly luncheon with his principal advisors.

Early in March 1968, air force SR-71 aircraft began arriving at Kadena to take over the BLACK SHIELD commitment. In progressive stages, the A-12 became standby to back up the SR-71. The last operational mission flown by OXCART occurred on 8 May 1968 over North Korea. Following this mission, the Kadena Detachment prepared its return home to Area 51.

Project headquarters selected 8 June 1968 as the earliest possible date to begin redeployment, and in the meantime limited flights of A-12 aircraft to those essential for maintaining flying safety and pilot proficiency. After the two surviving BLACK SHIELD aircraft had arrived in the US, they joined the others already in storage by the 7 June 1968 deadline.

The BLACK SHIELD pilots were briefed on the demise of the OXCART Program a few months before the actual shutdown of operations at the Kadena Detachment where the CIA kept the three A-12s as backup resources should the SR-71 not make its debut as advertised. Though the A-12s did backup the SR-71 missions, none were required to fly another mission. With about a month to go at Kadena, the A-12s were being prepped for their return to Area 51 and onto storage at the Lockheed Site 2 in Palmdale, California.

At the beginning of June 1968, the final month before the retirement of the A-12s, Jack Weeks and Frank Murray were the duty pilots at the detachment in Kadena. While awaiting the order to relocate, the three A-12s flew about once a week to provide currency for the pilots and to exercise the airplanes.

Each of the A-12s received a thorough final test flight to ensure its readiness for the flight over the Pacific to the home base at Area 51. A-12-06932, Article 129 as Lockheed called it was scheduled for its(FCF) on 3 June 1968. Frank Murray flew the functional check flight along the usual test hop route to the south of Kadena with a sweeping turn back to the north near the big island of the Philippines. During his flight, Murray discovered some problems with the scheduling of the internal bypass routine of the engine, leading to a flameout of the right engine on the descent back into Kadena. The difficulty with the engine was thought best solved by changing out the right engine with a "good" known engine kept at Kadena as a spare mission engine. Following an engine change, Jack Weeks was scheduled to fly another functional check flight on 5 June 1968.

The Jack Weeks Tragedy

The CIA A-12 overflights of denied territory were over, Operation BLACK SHIELD finished, and everyone headed home with the program was struck a devastating blow.

Article 129 had undergone an engine change before the long flight to Palmdale. On 4 June 1968, Aircraft 129 departed Kadena Air Base on a required functional check flight following an engine change. It did not return.

Weeks' mission started off with a top-off refueling, then the test hop route dictated by weather conditions. The climb-out and start of the turn back north apparently went as planned. Then the onboard telemetry system (Birdwatcher) started sending downlinks indicating the right engine was overtemping, followed by low fuel flow signals and finally that the altitude was below 70,000 approximately. Then all signals from Article 129 ceased, and everyone presumed the airplane was down. Intensive searches failed to show any sign of the aircraft or its pilot. The ensuing accident board concluded that the most likely cause of the loss of the plane was a catastrophic failure of the right engine. About this time, Article 127 and Article 131 were test flown and readied for their ferry back to Area 51. At Area 51, the commander, Colonel Slater sent two other A-12 pilots, Ken Collins, and Dennis Sullivan to Kadena to help with the ferry flights.

With no visual or radar sightings reported, the last known position of the aircraft was 520 nautical miles east of Manila. Search and rescue operations were begun shortly after the mishap.

Though the primary cause of the loss remained undetermined, the most probable cause was a catastrophic failure of an engine.

In addition to the five A-12s lost during the OXCART Program, the program lost two F-101 aircraft with both accidents fatal to the pilots.

On 1 June 1967, Lt Col Welton King left Kadena on a weather flight when the aircraft tail section separated in flight shortly after take-off. He safely ejected from the F-101 but died with the plane crashed on top of him.

On 26 September 19.67, Lt Col James S. Simon, Jr, perished when he inadvertently flew his aircraft into the ground at Area 51.

The two remaining A-I2s redeployed to Area 51 in June 1968. After post-flight maintenance, the pilots ferried them to Palmdale to join the rest of the OXCART fleet in storage.

The Kadena operation closed out as of 30 June 1968. Of the total thirteen A-12s procured, eight remained. Aircraft installed and ground equipment (with spare systems) kept at Palmdale supported all the A-12s in storage. Spare parts and equipment were stored at Palmdale to support at least a 90-day level for the five operational A-12 aircraft stored. It became possible to remove the plane from storage at some future date and prepare them for operation. Returning them to flight status was possible, but at a high cost in money and time. (In 1991, the air force released the A-12 planes to museums. Moving them by ground transport required cutting off the wings, which rendered them unflyable. On one, test Article 121 remains flyable in that regards as it never left the Palmdale area. The Flight Test Museum towed it to its retirement at the Blackbird Park in Palmdale with its wings intact.)

Jack Weeks

Headed Home to Area 51

Two A-12s remained after the loss of Jack Week's airplane. Dennis Sullivan flew one of the A-12s home to Area 51 without a hitch. The last plane launched with Ken Collins in the cockpit. He discovered a fuel leak that forced him to make an unplanned landing at Wake Island, where recovery crew repaired the fuel leak in a few days. Collins's tanker landed with him, so he refueled on the ground and flew a low/slow buddy flight with the tanker to Hawaii.

Frank Murray was the spare pilot, riding along with the tanker. After Collins had landed at Hawaii, he flew home to attend Jack Weeks memorial service while Murray stayed on at Hawaii to fly the final A-12 back to Area 51 with the same problem of a broken fuel manifold forcing repeated launches for home. Lockheed finally changed the entire engine accessory box, and the airplane flew on home to Area 51.

CIA A-12 pilot took a lot of ribbing after experiencing aircraft problems as he returned his Mach 3 A-12 from Kadena to Area 51. CIA via TD Barnes Photo Collection

Final Flight of the A-12

The earlier A-12's were ferried to Palmdale in the dark early morning hours not to be noticed by the public. Two days after the last A-12 landed at Area, 51 the relevant parties sanitized it of sensitive equipment and readied it for flight into storage at Palmdale to join the other A-12s already in storage. The morning of 21 June 1968, Frank Murray was scheduled to fly Article 131, the last A-12, to Palmdale to join the others in retirement. Frank considered this to be a singular honor even though one might say he won the honor by default since he was the only A-12 pilot left at the Area. To him, this was both a happy day and a sad one. He was glad to be the guy to fly the last flight and sad to see the end of the OXCART Program.

The Final Briefing:

The briefing was done about 0400 on the morning of 21 June 1968. Most personnel at Area 51 now of day were asleep, but as usual, some worked these hours to get things done, like launching the final A-12 mission. The mission was short with the routing over unpopulated areas and flown at subsonic speeds to prevent "booming" folks along the way.

The weather officer on duty for this briefing was none other than Major Willie Wuest - The Bard of Beatty. He wrote another of his fabulous oldies, appropriate for this occasion. In the words of the Bard himself.

"Composed and read at that tragic and historic occasion, the Final Scope Barn, 21 June 1968, in the Fifth Year of the Great Society.
The Last Briefing
With infinite skill and loving care

And a sad and heavy heart,
The weatherman in the early Morn
Drew his final weather chart.
He lovingly sketched each isobar
Over land and sea and shore,
As he had done so many times
For the last eight years or more.
He entered weather symbols
Each in its proper place,
As emotion overcame him
And tears streamed down his face.
His voice so choked with sorrow
You could barely hear him say
"This briefing chart is all screwed up,
It CAVU all the way."

After the Final Briefing, Murray stopped by the Command Post to bid farewell to the Duty Crew. It was there that Frank picked up the miniature A-12 metal marker the CP people used to mark the position of a A-12 flying out of the local area. This marker had the number 131 engraved on it.

Then it was off to the Personal Equipment room to don the pressure suit for the flight to Palmdale with farewells to the technicians who performed suit up of the pilot and checkout of his systems before flight. Most of these technicians were from the maker of the suits, the David Clark Company. With the suit checking OK, it time to ride the crew bus down to the hangar where 131 was waiting. Again, Frank paid farewells to the launch crewmen and the assembled few people there to see the final Launch.

The sound of the Buicks on the start cart still sounded like music; there was no sound like the roar of those engines as they turned the J-58 to start speed.

It was still dark on the taxi-out to Runway 32 where the runway launch crew chocked the A-12 for the pre-launch run-up. Here would be no chase on this final flight.

After takeoff, Frank made a right turn to the northeast and climbed up to about 40,000, where he turned back towards the Area, accelerating in AB to supersonic, leaving a wake-up call for the sleepy heads not involved with the final flight. He decelerated and took up the INS route to Palmdale, checking in with Salt Lake Center as he flew to Palmdale at about 30,000' foot altitude. Descending en route to the field at Palmdale, he checked with the tower for landing instructions.

The other project pilots and their families, along with Sharlene Weeks and her family could view the final flight and recovery into the storage area at Lockheed Site 2. Murray made a couple of flybys, one with a climb with afterburner after a low approach. This was the first and only time the family member saw the A-12 in flight. One of Murray's daughters missed the event since she was away at Girl Scout camp.

After landing, Murray taxied to the area where the families were assembled, then shut down Article 131 for the last time.

Murray exited the aircraft and unsuited, before joining the assembled families for a drive to the Antelope Valley Inn for breakfast.

The final flight was complete; the pilots left the CIA and rejoined the Air Force to complete their military careers. The Air Force rescinded their resignations during their sheep-dipping to join the CIA, giving the pilots credit for their time in the CIA as though they never resigned their commissions.

Postscript Project OXCART

In summary; the OXCART Program lasted just over ten years, from its inception in 1957 through first flights in 1962 to termination in 1968. It flew 2,850 flights out of Area 51. Lockheed produced 15 A-12s, three YF-12As, and 31 SR-71s. The 49 supersonic aircraft had completed more than 7,300 flights, with

17,000 hours in the air. Over 2,400 hours were above Mach 3. During the training phase and the Operation BLACK SHIELD phase combined, Project OXCART lost five A-12s in accidents with two pilots killed, and two experiencing narrow escapes. Also, the project lost two F-101 chase planes with their air force pilots.

Operation BLACK SHIELD demonstrated another of Area 51's power projection projects, this one being the CIA flying surveillance missions over North Korea and North Vietnam while supporting the United States at war.

The CIA achieved its main objective of creating a reconnaissance aircraft of unprecedented speed, range, and altitude capability.

The most important aspects of the effort lay in its byproducts, the notable advances in aerodynamic design, engine performance, cameras, electronic countermeasures, pilot life support systems, antiair devices, and above all in milling, machining, and shaping titanium. Altogether it was a pioneering accomplishment.

In a ceremony at Area 51 on 26 June 1968, Vice Admiral Rufus L. Taylor, deputy director of Central Intelligence, presented the CIA Intelligence Star for Valor to pilots Kenneth S. Collins, Ronald L. Layton, Francis J. Murray, Dennis B. Sullivan, and Mele Vojvodich for participation in the BLACK SHIELD operation. Diane, his widow, accepted the posthumous award to pilot Jack W. Weeks.

At Area 51, the air force presented the Air Force Legion of Merit to Colonel Slater and his deputy, Colonel Maynard N. Amundson. At the same event, it awarded the Air Force Outstanding Unit Award to the members of the OXCART Detachment (1129th Special Activities Squadron, Detachment 1) and the air force supporting units for their achievements during project BLACK SHIELD's contribution to the Vietnam War. (Note: The air force activated the 1129th SAS strictly to support Project OXCART at Area 51. The unit disbanded at the end of Project OXCART at Area 51.)

Spouses of the pilots attended the presentation and learned for the first time of the activities of their wives. Kelly Johnson spoke as a guest speaker at the ceremony and lamented in moving words the end of an enterprise marking his most outstanding achievement in aircraft design. He received his awards the president's Medal of Freedom in 1964 and on 10 February 1966, the National Medal of Science, from President Johnson, for his contributions to aerospace science and the national security.

At CIA headquarters in Langley, Virginia, the Memorial Wall contains two stars honoring the two CIA A-12 pilots who died in plane crashes - Walter L. Ray (died January 5, 1967,), and Jack W. Weeks (died June 4, 1968). The CIA honored both in them the CIA Book of Honor.

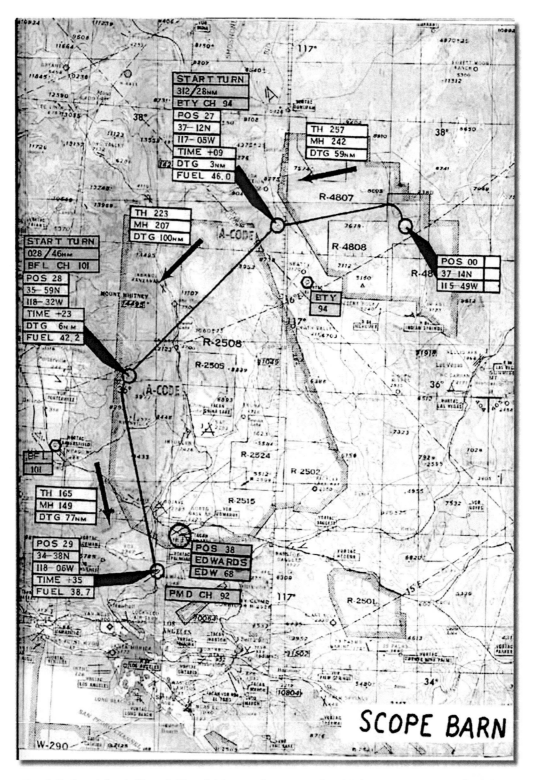

Final flight of the A-12 with Frank Murray flying Article 131 from Area 51 to Palmdale for storage. Frank Murray

On 26 June 1968, the CIA Intelligence Star for Valor was awarded to CIA Oxcart and Black Shield Mission Pilots (L to R): Mel Vojvodich, Dennis Sullivan, Jack Layton, Ken Collins, and Frank Murray by Deputy Director of the CIA V.Adm. Rufus Taylor. The widow of Jack Weeks posthumously accepted his award. CIA via TD Barnes Photo Collection

The surviving A-12s in storage at Palmdale, California where they remained in secret from June 1958 to January 1991.

Boots on the Ground at Area 51

Most of the people working on the projects at Area 51 never set foot there. Many working on Area 51 projects do so unknowingly. Declassification by the CIA over the years have revealed the identity of some. Only through their membership in the Roadrunners Internationale social association of those who worked at Area 51 for the CIA have the identities of some become known.

The Presidents
Harry S. Truman
Dwight D. Eisenhower
John F. Kennedy
Lyndon B. Johnson
Richard Nixon
Gerald Ford
Jimmy Carter
Ronald Reagan
George H. W. Bush
Bill Clinton
George W. Bush

The CIA Directors
Roscoe H. Hillenkoetter,	1947–1950
Walter Bedell Smith,	1950–1953
Allen W. Dulles	1953–1961
John McCone	1961–1965
William Raborn	1965–1966
Richard M. Helms	1966–1973
James R. Schlesinger	1973
William Colby	1973–1976
George H. W. Bush	1976–1977
Stansfield Turner	1977–1981
William J. Casey	1981–1987
William H. Webster	1987–1991
Robert M. Gates	1991–1993
R. James Woolsey	1993–1995
John M. Deutch	1995–1996
George J. Tenet	1996–2004
Porter J. Goss	2004–2005

CIA commanders of Area 51
Richard A. "Dick" Newton USMCR (Ret) (1955–1956)
Landon McConnell (1956–1957)
Werner Weiss (1958–1961)
Richard A. "Dick" Sampson (1969–1971)
Sam Mitchell (1971–1977)

Air Force Commanders of Area 51 During the CIA Era.
Col William R. Yancey, commander of the 4070th SAS during CIA U-2 Project AQUATONE, retired a brigadier general with CIA's Newton, and then McConnell as deputy

USAF Col Robert J. Holbury, commander of the 1129th SAS (December 1961 to July 1966) with CIA's Werner Weiss as deputy

USAF Col Hugh "Slippery" Slater, commander of the 1129th SAS (July 1966 to 1969) with Werner Weiss as deputy

USAF Col Larry D. McClain (1977 to 1 April 1979) first USAF site manager/detachment commander

The Honorable Gene Poteat, retired CIA for his technical contributions to Area 51. Poteat fathered information warfare. With Project PALLADIUM, he forever changes the course of US aerial reconnaissance.

Helen H. Kleyla who compiled the report declassified by the CIA and herein reference by the author. The declassified report documents the political turmoil leading to the Central Intelligence Agency establishing Area 51. Kleyla served the CIA for 30-years, much of it as the assistant to CIA's Dick Bissell. Kleyla was the first woman to set foot at the CIA facility in Area 51.

Richard Mervin Bissell, Jr., the Central Intelligence Agency officer responsible for the U-2 spy plane. Bissell and Herbert Miller, another CIA officer, chose Area 51 as the CIA's test facility. As the CIA's DD/P, Deputy Director of Plans, Bissell also oversaw the early stages of Project OXCART, the development of the Lockheed A-12.

Dr. Albert D. "Bud" Wheelon, Ph.D., the first CIA DDST for his support of Project OXCART.

CIA's John Parangosky, project manager for OXCART for developed America's first stealth plane that today remains the fastest, highest-flying, pilot jet aircraft ever.

. Richard A. "Dick" Newton USMCR (Ret) (1955–1956) is the first CIA commander at the Groom Lake facility as Watertown, followed by Landon McConnell (1956–1957), Richard A. "Dick" Newton USMCR (Ret) (1955–1956), and Landon McConnell (1956–1957). When the CIA reopened Area 51 as the Paradise Ranch for Project OXCART during 1961, the CIA appointed Werner Weiss, who remained until 1969, at which time Richard A. "Dick" Sampson, who remained until 1971. Sam Mitchell was the CIA commander from 1971 to 1980 when Col Larry McClain with the US Air Force assumed control in late 1979.

USAF Col William Yancey is the first Air Force commander at Area 51, commanding the 4070th Support Group with the CIA's Newton. Yancey's deputy, Col Landon McConnell follows during 1956.

During December 1961, Col Robert J. Holbury, Call Sign: Dutch 10, assumed command of Detachment 1 of the 1129th US Air Force Special Activities Squadron for the A-12 Project OXCART. He served until July 1966, with the CIA's Werner Weiss as his deputy.

USAF Col Hugh "Slippery" Slater takes command of the 1129th SAS during July 1966, and remains until 1969, with Werner Weiss remaining as his deputy. Col Larry D. McClain assumes command during 1979 as the first USAF site manager/detachment commander. Brig. Gen William R. Yancey-Project AQUATONE, Brig. Gen Robert J. Holbury-Dutch 10-Project OXCART, Col Hugh "Slip" Slater-Dutch 11

Col Hugh Slater, born in Seattle, Washington, grows up in Los Angles where he attended USC. During World War II, he flew 84 combat missions and supported the Berlin airlift. He first worked for the CIA during March 1963, with a classified assignment to the Chinese 35th Black Cat, U-2 Squadron in Taiwan, where he checked out in the U-2 and flew four flights at North Base, Edwards Air Force Base. He continued to serve the CIA at Area 51 as the operations officer under Col Robert Holbury and then as the commander for two more years including Operation BLACK SHIELD. He received the Distinguished Service Medal, the Legion of Merit, and the Distinguished Flying Cross with a cluster and the Air Medal with 12 clusters. In addition, he received the CIA Medal of Merit.

Col Hugh "Slip" Slater - Area 51 Commander Project OXCART/OPERATION BLACK SHIELD

Photographed w/YF-12

Col Am Amundson

The famous Murphy Green is setting up for celebrating the 500th flight of the A-12 trainer known as the Titanium Goose. Major Culp is on the left and Colonel Am Amundson on the right.

L-R front row: Burt Barrett, 1129th SAS Commander Col Slip Slater Col Amundson

RICHARD A. SAMPSON

Richard A. "Dick" Sampson (CIA), born during 1927, married with two children, joined Project OXCART during 1964, when assigned to the Services and Support Group, USAF Headquarters Command at Bolling Air Force Base, DC, where he works with Joe Murphy and Herb Saunders to handle project security for the IDEALIST (U-2), OXCART (A-12), and SENIOR CROWN (SR-71) programs.

The SR-71 HABU

The air force was exceedingly helpful all through the CIA's OXCART program, providing financial support, conducting the refueling program, providing operational facilities at Kadena, and airlifting OXCART personnel and supplies to Okinawa for the operations over Vietnam and North Korea. The air force also ordered a small fleet of A-11's, which Lockheed produced as two seated reconnaissance aircraft named the SR-71. The SR-71 replacements for the A-12 became operational about 1967.

The stated mission of the SR-71 conducted a "post-strike reconnaissance," that is, looking at the enemy situation after a nuclear exchange. The likelihood of using the aircraft in the capacity hardly appeared significant. However, the SR-71 remained capable of conducting regular intelligence missions.

For these purposes, however, the A-12 possessed certain definite advantages. The A-12 carried only one man, leaving more room for cameras, as well as for various other collection devices, which at the time the SR-71 could not carry. It was certainly the most effective reconnaissance aircraft in existence, or likely to become in existence for years to come. Operated by civilians, it covertly deployed without the number of personnel and amount of fanfare typically attending an air force operation.

The air force's ordering the SR-71s eased the path of OXCART development. It shared the financial burden and reduced somewhat the cost per aircraft by producing greater numbers. In the longer run, however, the existence of SR-71 spelled the doom of OXCART, the reasons chiefly financial and the extent of the related capabilities of A-12 and SR-71.

With the CIA now out of the aerial reconnaissance business, its political foes directed their attention on the air force's SR-71. The SR-71 also was not liked by many of the "overhead" or satellite people at the NRO and elsewhere who thought satellites were the best for everything. They considered the SR-71 to be competitive, rather than the complementary system. By their very nature, satellite successes (and failures) tended to be very hush-hush and here was the SR-71 getting the "glory." Further, there was the oft-stated opinion by many that satellites did anything needed better than anything else. While they are marvelous devices, this has never been true. The SR-71 flying around tended to gainsay the omnipotent image of satellites.

In the 1980s, opposition to the SR operations got stronger in some areas, canceling planned sensor and maintainability upgrades. Then, because the SR could not perform the function the upgrade was supposed to do, this became a reason the SR was not capable enough. To cite some examples much was made of the SR-71's inability to downlink its data to commanders in the field and returning to base for processing its results. The air force developed and flight tested a data link on the SR-71 in 1980, and then ordered the system installed on the TR-1, but not the SR-71.

Secretary of Defense Cheney did not like the plane while a congressman and wanted to kill the SR-71, despite Congress' willingness to continue funding the program.

Robert McNamara also hated the SR-71, and in one of his series of colossally stupid moves, ordered all tooling destroyed in the late 1960s to kill the program for lack of spare parts. Lockheed destroyed the tooling, but not before manufacturing and producing a twenty-year supply of spare parts, which included $60,000,000 of Russian titanium sheets. The air force moved the spare parts from Norton AFB, California for storage in three Marine Corps logistics warehouses in Barstow, California.

The SR-71 served with the air force from 1964 to 1998. Of 32 aircraft built, the air force lost 12 in accidents with none lost to enemy action. NASA used three SR-71 aircraft at different times during the 1990s as test beds for high-speed and high-altitude aeronautical research.

The last of the SR-71s found a home in various museums. However, the spare parts remained in the Marine Corps warehouses, costing a fortune to store hundreds of spare tires, plane parts containing asbestos, requiring special care to meet environmental regulations. The Marine Corps needed its warehouses to support the Iraqi War and asked the air force to remove its spare parts. The air force denied ownership, claiming that NASA took ownership when it took three SR-71s for research flight.

NASA also denied ownership, and the matter eventually went to court to determine ownership and liability. The air force lost the court case and disposed of the spare parts by having them scrapped by a metal scrap dealer in Tucson, Arizona. Sadly, this included new Buick engines for use as starters for the SR-71s and the sheets of titanium valued at millions of dollars.

With the nation's SIGINT activities now consolidated under NSA auspices, NSA and the air force carried on and expanded the United States' special collection programs.

Meanwhile, the work continued at Area 51 for many of those participating in Project AQUATONE, Project OXCART, and Operation BLACK SHIELD. As the Blackbirds moved out, the MiG-21 moved in for project DOUGHNUT.

In January 1968, project HAVE DOUGHNUT, a joint Air Force/Navy technical and tactical evaluation of the MiG-21F-13 began at Area 51 at the same time the Blackbird family moved out.

BLACKBIRD FAMILY

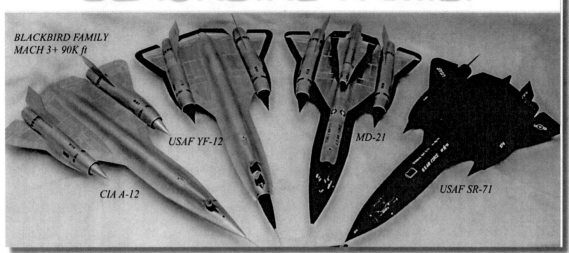

Blackbird Family Losses

60-6926: A-12

This was the second A-12 to fly but the first to crash. On 24 May 1963, CIA pilot Ken Collins was flying an inertial navigation system test mission. After entering clouds, frozen water fouled the pitot-static boom and prevented correct information from reaching the standby flight instruments and the Triple Display Indicator. The aircraft subsequently entered a stall and control was lost completely followed by the onset of an inverted flat spin. The pilot ejected safely. The wreckage was recovered in two days. Persons at the scene were identified and requested to sign secrecy agreements. A cover story for the press described the accident as occurring to an F-105 and still remains listed in this way on official records.

60-6928: A-12

This aircraft was lost on 5 January 1967 during a training sortie flown from Groom Lake. Following the onset of a fuel emergency caused by a failing fuel gauge, the aircraft ran out of fuel only minutes before landing. CIA pilot Walter Ray was forced to eject. Unfortunately, the ejection seat man-seat separation sequence malfunctioned, and Ray was killed on impact with the ground, still strapped to his seat.

60-6929: A-12

This aircraft was lost on 28 December 1967 seven seconds into an FCF (Functional Check Flight) from Groom Lake performed by CIA pilot Mel Vojvodich. The SAS (Stability Augmentation System) had been incorrectly wired up, and the pilot was unable to control the aircraft 100 feet above the runway. The pilot ejected safely. A similar accident occurred when the first production Lockheed F-117 was flown on 20 April 1982 by Bill Park. Its control system was hooked up incorrectly. Bill Park survived the accident but had injuries serious enough to remove him from flight status.

60-6932: A-12

This aircraft was lost in the South China Sea on 5 June 1968. CIA pilot Jack Weeks was flying what was to be the last operational A-12 mission from the overseas A-12 base at Kadena AB, Okinawa. The loss was due to an inflight emergency, and the pilot did not survive. Once again, the official news release identified the lost aircraft as an SR-71 and security was maintained. A few days later the two remaining planes on Okinawa flew to the US and were stored with the remainder of the OXCART family.

60-6934: YF-12A

This aircraft, the 1st YF-12A, was seriously damaged on 14 August 1966 during a landing accident at Edwards AFB. The rear half was later used to build the SR-71C (64-17981) which flew for the first time on March 14, 1969.

60-6936: YF-12A

This aircraft, the third YF-12A, was lost on 24 June 1971 in an accident at Edwards AFB. Lt. Col. Ronald J. Layton and systems operator William A. Curtis were approaching the traffic pattern when a fire broke out due to a fuel line fracture caused by metal fatigue. The flames quickly enveloped the entire aircraft and on the base leg, both crew members ejected. '936 was destroyed.

60-6939: A-12

This aircraft was lost on approach to Groom Lake on 9 July 1964 following a Mach 3 check flight. On approach, the flight controls locked up, and Lockheed test pilot Bill Park was forced to eject at an altitude of 200 feet in a 45-degree bank angle!

60-6941: M-21

This was the second A-12 converted to an M-21 for launching the D-21 reconnaissance drone. During a flight test on 30 July 1966 for launching the drone, the drone pitched down and struck the M-21, breaking it in half. Pilot Bill Park and LCO (Launch Control Officer) Ray Torrick stayed with the plane a short time before ejecting over the Pacific Ocean. Both made safe ejections, but Ray Torrick opened his helmet visor by mistake, and his suit filled up with water which caused him to drown. This terrible personal and professional loss drove "Kelly" Johnson to cancel the M-21/D-21 program.

64-17950: SR-71A

The prototype SR-71 was lost on 10 January 1967 at Edwards during an anti-skid braking system evaluation. The main undercarriage tires blew out, and the resulting fire in the magnesium wheels spread to the rest of the aircraft as it ran off the end of the runway. Lockheed test pilot Art Peterson survived.

64-17952: SR-71A

This aircraft disintegrated on 25 January 1966 during a high-speed, high-altitude test flight when it developed a severe case of engine unstart. Lockheed test pilot Bill Weaver survived although his ejection seat never left the plane! RSO (Reconnaissance System Officer) Jim Zwayer died in a high-G bailout. The incident occurred near Tucumcari, New Mexico.

64-17953: SR-71A

This aircraft was lost on 18 December 1969 after an inflight explosion and subsequent high-speed stall. Lt. Col. Joe Rogers and RSO Lt. Col. Garry Heidelbaugh ejected safely. The precise cause of the explosion has never been determined. The incident occurred near Shoshone, California.

64-17954: SR-71A

This aircraft was demolished on 11 April 1969 under the circumstances like 64-17950. New aluminum wheels and stronger tires with beefed up compound were retrofitted to all SR-71s. Lt. Col. Bill Skliar and his RSO Major Noel Warner managed to escape uninjured.

64-17957: SR-71B

This aircraft was the second SR-71B built for the Air Force. It crashed on approach to Beale on 11 January 1968 when instructor pilot Lt. Col. Robert G. Sowers and his "student" Captain David E. Fruehauf were forced to eject about 7 miles from Beale after losing all control. The plane had suffered a double generator failure followed by a double flameout (caused by fuel cavitation) and pancaked upside down in a farmer's field.

64-17965: SR-71A

This aircraft was lost on 25 October 1967 after an INS platform failed, leading to the display of erroneous attitude information in the cockpit. During a night flight, the INS gyro had tumbled. There were no warning lights to alert pilot Captain Roy L. St. Martin and RSO Captain John F. Carnochan. In total darkness, in a steep dive and no external visual references available, the crew had little alternative. They ejected safely near Lovelock, Nevada.

64-17966: SR-71A

Lost on the evening of 13 April 1967 after the aircraft entered a subsonic, high-speed stall. Pilot Captain Earle M. Boone and RSO Captain Richard E. Sheffield ejected safely. The incident occurred near Las Vegas, Nevada.

64-17969: SR-71A

Lost on 10 May 1970 during an operational mission, from Kadena, against North Vietnam. Shortly after air-refueling, the pilot, Major William E. Lawson initiated a normal full power climb. Stretching before him was a solid bank of cloud containing heavy thunderstorm activity which reached above 45,000'. Heavy with fuel, the aircraft was unable to maintain a high rate of climb, and as it entered turbulence both engines flamed out. The rpm dropped to a level too low for restarting the engines. Pilot and RSO, Major Gilbert Martinez ejected safely after the aircraft stalled. The plane crashed near Korat RTAFB, Thailand.

64-17970: SR-71A

Lost on 17 June 1970 following a post-tanking collision with the KC-135 tanker. Lt. Col. "Buddy" L. Brown and his RSO Maj. Mortimer Jarvis ejected safely although both legs of the pilot were broken. The SR-71 crashed 20 miles east of El Paso, Texas, but the KC-135 limped back to Beale AFB with a damaged fin.

64-17974: SR-71A

This aircraft was lost on 21 April 1989 over the South China Sea and is the last loss of any Blackbird as of December 1991. Pilot Lt. Col. Dan House said the left engine blew up and shrapnel from it hit the right-side hydraulic lines, causing a loss of flight controls. House and RSO Blair Bozek ejected and came down safely in the ocean. They had been able to broadcast their position before abandoning the Blackbird, and rescue forces were immediately on the way. However, the crew were rescued by native fisherman. The local chieftain's new throne is Colonel House's ejection seat!

64-17977: SR-71A

This aircraft ended its career in flames by skidding 1000 feet off the end of runway 14 at Beale on 10 October 1968. The takeoff was aborted when a wheel assembly failed. Major James A. Kogler was ordered to eject, but pilot Major Gabriel Kardong elected to stay with the aircraft. Both officers survived.

64-17978: SR-71A

Nicknamed the Rapid Rabbit, this aircraft was written off on 20 July 1972 during the rollout phase of its landing. The pilot, Captain Dennis K. Bush, had practiced a rapid deploy-jettison of the braking parachute. A go-around was initiated after the chute was jettisoned. On the next landing attempt, the aircraft touched down slightly "hot" but had no chute to reduce the aircraft's speed. The pilot was unable to keep the plane on the runway. A wheel truck hit a concrete barrier. The aircraft suffered significant damage. The pilot and the RSO, Captain James W. Fagg, escaped without injury.

A total of 20 Lockheed Blackbirds were lost due to a variety of accidents; however, not one was shot down by unfriendly forces!

Stats

Broken down by type:

Aircraft designation:	A-12	M-21	YF-12	SR-71A	SR-71B	SR-71C
No. of aircraft built:	13	2	3	29	2	1
No. of aircraft lost:	5	1	2*	11	1	0

* SR-71C (64-17981) was built using the rear half of YF-12A (60-6934) and functional engineering mockup of the SR-71A forward fuselage.

Written by Al Dobyns Maintained by Carl Pettypiece

Chapter 13 - Summary

On 19 and 20 July 1967, the CIA achieved a new high-water mark in aerial reconnaissance when the CIA flowed three photographic/ELINT reconnaissance sorties in the A-12 and U-2 aircraft over South China and North Vietnam territory. These were A-12 missions BX6709, 19 July Zulu and BX6710, 20 July Zulu, and U-2 mission C237C, 20 July Zulu. The first of the sorties, BX6709, entered denied territory at 0424Z/19 July and the last of the sorties, C237C, exited denied area at 0632/20 July, a total elapsed time of only 26 hours and 08 minutes from first entry to the last exit. Thus, the CIA significantly accomplished first back-to-back A-12 sorties, a first simultaneous overflight by A-12 and U-2 aircraft. Also, these sorties photographed numerous tactical targets such as power plants, industrial complexes, rail-heads/lines, barracks areas, warehouse areas, and new construction areas. The significance of this was the CIA's capability to provide high-resolution wide-swath photography quickly responsive to weather patterns. Also, it was an actual demonstration that the A-12 was the only low-vulnerability manned reconnaissance vehicle capable of providing such photography over a SAM-defended area.

The CIA alerted twenty-six BLACK SHIELD high-altitude reconnaissance missions during the period from 16 August to 31 December 1967. Fifteen of the 26 missions flown with two fired at by enemy surface-to-air missile units. Eleven missions canceled, due to weather conditions.

Through 31 December 1967, BLACK SHIELD coverage showed no evidence of surface-to-surface missiles, equipment, or facilities. The coverage continued to be of special value for investigating reports of missile deployment derived from other sources. For example, coverage of the Haiphong harbor area obtained by BX6723 followed the arrival there of the Soviet freighter Partizanskaya Iskra, which a clandestine source reported as carrying SSN-2 (Styx) cruise missile crates. A search of the ship and port area photography did not reveal any such crates.

The BLACK SHIELD program obtained good baseline coverage. There now existed clear, interpretable photography of all North Vietnam except for a small area next to the Chinese border in Northeast North Vietnam. BLACK SHIELD photography continued to be invaluable in offering a unique order of battle information on fighter aircraft and surface-to-air missiles.

Some missions provided total coverage of most of the major North Vietnamese airfields. The simultaneous coverage of many surface-to-air missile sites gave US theater forces a quick, comprehensive listing of the occupied SA-2 sites. It also significantly supplemented communications intelligence in finding the actual number of surface-to-air missile battalions in North Vietnam and aided in predicting with some degree of certainty the SA-2 occupancy status for tomorrow's strike operation. As an example, back-to-back missions BX6739 and BX6740, flown 15 and 16 December 1967, photographed 221 of North Vietnam's 226 useable SA-2 sites, including six new locations. The mission found twenty of these SA-2 sites occupied, including five of the six new locations.

The program continued to contribute to bomb damage assessment of point targets and the interdiction effort directed against North Vietnam's road, rail, and water transportation systems. BLACK SHIELD missions identified new targets for US air strikes and offered valuable information on Chinese military activity, both in North Vietnam and along the southern and: western coast of Hainan Island. Also, BLACK SHIELD missions BX6737 and BX6738 were flown over the border areas of Cambodia, Laos, and South Vietnam to obtain information on the North Vietnamese infiltration and supply routes and the major North Vietnamese-Viet Cong troop deployment zones.

All the North Vietnamese missions, as well as Cambodian BLACK SHIELD mission BX6737 (which extended up the North Vietnamese panhandle), were traced by Chinese and North Vietnamese air surveillance facilities. North Vietnam's air defense reaction to the vehicle first occurred in related surface-to-air missile communications traffic during mission BX6723 of 17 September. Subsequently, Mission

BX6727 of 6 October confirmed surveillance tracking of the vehicle by North Vietnamese facilities.

On 6 October 1967, Frank Murray flew Mission BX6727 that established air surveillance tracking of the vehicle by North Vietnamese facilities. BLACK SHIELD protective mechanisms activated on missions BX6716, BX6725, BX6732, and BX6734 with one surface-to-air missile launched at the A-12 on Dennis Sullivan's mission BX6732. The enemy fired nine to eleven SAMs at BLACK SHIELD mission BX6734.

The BLACK SHIELD surveillance program saw a change in the North Vietnamese surface-to-air missile launch tactics on missions BX6732 and BX6734. The first three missile firings occurred with the Fan Song guidance radar in low PRF (pulse repetition frequency). To cope with a Mach-3 target use of the low PRF mode permits an earlier missile launch against very fast targets. North Vietnam air defense facilities were forewarned of the 6 October and 5 October missions (BX6727 and BX6728), apparently because of intercepting transmissions from the South China Sea refueling area. A more secure communications posture since employed noted no more forewarning messages.

BLACK SHIELD Operational Missions Alerted Between 31 May and 15 August 1967

Mission No.	Date (1967)	Pilot	Remarks
BSX-001	31 May	Vojvodich	Flown
BSX-002	6 June		Canceled due to weather
BSX-003	10 June	Weeks	Flown
BX6704	10 June		Cancelled due to weather
BX6705	20 June	Layton	Flown
BX6706	30 June	Weeks	Flown
BX6707	30 June		Cancelled due to weather
BX6708	13 July	Collins	Flown (Rt. 19, modified)
BX6709	19 July	Sullivan	Flown (Rt. 9, modified)
BX6710	20 July	Collins	Flown (Rt. 14, modified)
BX6711	29 July		Cancelled due to weather
BX6712	30 July		Cancelled due to weather
BX6713	13 August		Cancelled due to weather
BX6714	14 August		Cancelled due to weather
BX6715	14 August		Cancelled due to weather

A-12 Inventory

Aircraft Number	SN	Configuration	Number of Flights	Hours Flown	Disposition
121	60-6924	Flight testing	332	418.2	Blackbird Airpark, Palmdale, CA
122	60-6925	Systems/testing	122	177.9	USS Intrepid Sea-Air-Space Museum, New York, NY
123	60-6926	Operations	79	135.3	Crashed 24 May 1963
124	60-6927	Training	614	1076	California Science Center, Los Angeles, CA
125	60-6928	Operations	202	334.9	Crashed 5 January 1967
126	60-6929	Operations	105	169.2	Crashed 28 December 1965
127	60-6930	Operations	258	499	Alabama Space and Rocket Center, Huntsville, AL
128	60-6931	Operations	232	453	CIA headquarters, Langley, VA
129	60-6932	Operations	268	410	Crashed 4 June 1968
130	60-6933	Operations	217	406	San Diego Aerospace Museum, San Diego, CA

131	60-6937*	Operations	177	345.8	Southern Museum of Flight, Birmingham, AL
132	60-6938	Operations	197	370	USS Alabama Battleship Memorial Park, Mobile, AL
133	60-6939	Operations	10	8.3	Crashed 9 July 1964
134	60-6940	Drone operations	80	123.9	Museum of Flight, Seattle, WA
135	60-6941	Drone Operations	95	152	Crashed 30 July 1966

*Numbers 6934 to 6936 were used for the air force's YF-12A fighter-interceptor version.

OXCART Accomplishments.

The OXCART program lasted nine years.

The total cost, including development, production, maintenance and operation, arid support amounted to $950 million. From concept to operation, it pioneered the way in the aerospace industry to reach the plateau of Mach 3.2 flight.

The technical data accumulated over the nine years have made, and will continue to make, significant contributions to developments in supersonic aircraft, high Mach turbine engines, aerial reconnaissance, electronic countermeasures, life support and ancillary systems. It was responsive to priority intelligence requirements when maintaining surveillance of North Vietnam. It did not detect the introduction of offensive missiles but did provide bomb damage assessments, and overall military logistics estimate to the field commanders. It found the Pueblo in Wonsan Harbor and provided valuable information about the size and disposition of North Korean forces.

OXCART Phase-Out.

On 10 November 1965, Mr. W. R. Thomas, Chief of the International Division, and Mr. S. B. Leach, Chief of the Military Division, Bureau of the Budget, submitted a memorandum to the Budget Director. In it, they expressed concern at the total costs of the A-12 and SR-71 programs, both past and projected. They stated a cost of 2.5 billion dollars on both programs through FY 1966 and expected to pay $2.1 billion more through 1971. They questioned the need, first for the total number of aircraft represented in the combined fleets, and second, the need for a separate CIA (OXCART) fleet. Several alternatives offered to achieve a substantial reduction in forecast spending. They recommended phasing out the A-12 program by September 1966 and no further procurement of SR-71 aircraft. They distributed copies of this memorandum to the DOD, D/NRO, and DCI with the suggestion that these agencies explore the alternatives set out in the paper.

The secretary of defense declined even considering the proposal, presumably because the SR-71 would not be operational by September 1966.

The matter rested until July 1966 when Mr. Schultze, Director of the Budget, reopened the subject. He proposed that a study of the relationship between the OXCART and SR-71 programs by the DOD/CIA/BOB in time for FY 1968 budget deliberations. He suggested possible alternatives that the study group might examine:

1. Retention of separate A-12 and SR-71 fleets, i.e., status quo.
2. The co-location of the two fleets.
3. Transfer the OXCART mission and aircraft to Strategic Air Command.
4. Transfer the OXCART mission to Strategic Air Command and store the A-12s as attrition replacements for the SR-71s.
5. Transfer OXCART mission to Strategic Air Command and dispose of the A-12 aircraft. Schultze designated Mr. C. W. Fischer as the Bureau of Budget representative on the study group.

The DCI (Mr. Helms) appointed Mr. Carl Duckett, assistant: deputy director for Science and Technology, as the CIA member, and the Department of Defense named Mr. Herbert D. Bennington. Mr. Duckett shortly after that became the acting deputy director Science & Technology and was unable to devote the time needed for the study. He appointed Mr. John Parangosky, who was then the AD/OSA, as the CIA's member of the study group. During the summer and fall of 1966, The panel conducted a complete control system appraisal of the two fleets, examining the relevant technologies, operational capabilities, support facilities and costs and balanced the capabilities of advanced aircraft against those of satellites and drones. The review included the special covert and civilian characteristics of the OXCART fleet to anticipate the effect on U.S, relations in matters of clandestine reconnaissance in the event of the termination of the OXCART project.

The study group named three alternatives for such a decision. They were:

1. Maintain the status quo and continue both fleets at the currently approved levels. Estimated costs through EY 1972 would total $1.377 billion.

2. Mothball all A-12 aircraft, but maintain the OXCART capability by sharing SR-71 aircraft between Strategic Air Command and the CIA, saving $252 million over the first alternative.

3. Terminate the OXCART fleet in January 1968 {assuming an operational readiness date of September 1967 for the SR-71 and assign all missions to the SR-71 fleet. Adopting this alternative would save $365 million.

The report made no recommendations per se. Its purpose was to provide information for making higher level judgments. On 12 December 1966, Mr. Helms, Mr. Schultze, Mr. Vance, and Dr. Hornig, Scientific Advisor to the President attended a meeting at the Bureau of the Budget, where they voted on the alternatives posed in the Fischer Bennington Parangosky report. Messrs. Vance, Schultze and Hornig voted to terminate the OXCART fleet, and Mr. Helms voted for eventual sharing of the SR-71 fleet between CIA and Strategic Air Command.

The Bureau of Budget immediately prepared a letter to the president conveying the course of action recommended by the majority. Mr. Helms, having dissented from the majority, requested the deputy director Science & Technology sent the president a letter stating the case for CIA remaining in the reconnaissance business and his reasons for voting as he did. On 16 December 1966, he handed Mr. Helms a draft memorandum to the President requesting a decision to either to share the SR-71 fleet between CIA and Strategic Air Command or to terminate the CIA capability entirely.

On 20 December Mr. Helms wrote Mr. Schultze of his having new information of considerable significance concerning SR-71 performance. He requested another meeting after 1 January to review pertinent facts. He also asked for withholding the memorandum to the president pending that meeting's outcome. Specifically, obtained evidence and data indicated that the SR-71 program was having serious technical problems. No one doubted the SR-71 achieving an operational capability by the time the A-12 program terminated. Mr. Helms was so concerned about the SR-71's abilities, that he changed his position from sharing the SR-71 aircraft with Strategic Air Command to a firm recommendation to retain the OXCART fleet with wider private sponsorship, and separate basing. Other eleventh-hour attempts to review the subject were in vain.

On 28 December 1966, the president accepted the recommendations of Messrs. Vance, Hornig and Schultze, and directed the termination of the OXCART program by 1 January 1968.

The decision to terminate the OXCART program required, the development of an orderly phase-down procedure. After consultation with project headquarters, the director, National Reconnaissance Office advised the deputy secretary of defense on 10 January 1967 of the phase-out schedule of aircraft. The CIA would store four A-12s in July 1967, two more by December, and the last four by the end of January 1968. Until 1 July 1967, the OXCART Detachment was to maintain a global capability to conduct operational missions from a prepared overseas location and simultaneously from Area 51 in Nevada.

The operational readiness, posture was to include maintenance of a 15-day fast, reaction capability for deployment to the Far East and a 7~day quick response for implementation over Cuba. Between 1 July and

31 December 1967, the CIA would maintain the capability to conduct operational missions from either a prepared overseas base or from Area 51, but not simultaneously. The CIA would support a quick reaction capability for either Cuban overflights or deployment to the Far East.

On 9 May 1967, Secretary Vance directed that the SR-71 assume the responsibility of conducting Cuban flights as of 1 July 1967, and the dual capability of overflying Southeast Asia and Cuban by 1 December 1967. The director, National Reconnaissance Office established a joint CIA/air force working group to coordinate planning actions incident to the OXCART phase out. He gave the team the responsibility of identifying decisions and problem areas, recommending courses of action, and advising the NRO, CIA, and air force of progress in the phase-out.

The overall phase-out schedule operated under the code name SCOPE COTTON with the underlying assumptions of guiding the phase-out with no further significant updating of the A-12 aircraft. The group authorized only those engineering change proposals, retrofits, and modifications affected the safety of flight, or essential to maintain an operational capability. The CIA would mothball and store all aircraft by 31 January 1968, with no new construction approved at Area 51. The CIA would close Area 51 upon removal of planes and other assets. The CIA could not make new equipment procurements and could maintain, a level of spares and supplies at levels sufficient only to sustain the operational commitment through 1967.

The CIA would reduce the OXCART engine inventor, and make overhauls and improvements only as necessary. The working group suspended procurement actions for air force personnel, after which the director, National Reconnaissance Office identified all action items and issued a series of SCOPE COTTON decisions.

The TAGBOARD program would remain at Area 51 through 1967 along with OXCART assets at Kadena and deployment flyaway hits. The CIA would store twenty engines with the aircraft at Palmdale and redistribute repairable items more than OXCART requirements to other programs with no cannibalization of the A-12 airframes. The CIA would retain all guidance equipment, cameras. EWS and pilot equipment in minimum quantities and type to support the stored aircraft. The SR-71 would use the remaining stocks of conventional A-12 spares, and the air force would accept responsibility for follow-on contracting and funding for J58 engine product improvement as well as for the YF 12A, beginning in FY 1968. The CIA would redistribute any excess support assets to other National Reconnaissance Office programs' Overseas facilities except at Kadena, would: transfer to the base where located effective 1 January 1968. The air force made a gradual transfer of SR-71 contracting responsibilities.

The radar range at Area 51 would dismantle and move to air force installations. Research and development would suspend countermeasures as they applied to OXCART.

Project headquarters moved quickly to advise the contractors of the phase-out decision. At the same time, they, along with all witting personnel, received a caution to observe the same standards of security during termination as noted during the development and operational readiness stages of the program.

The security precaution Was necessary to protect the CIA's role, in the program, CIA procurement and contracting methods, and, most importantly, the mission posture for the year. The CIA withheld dissemination of the information even in the cleared community until it worked out an orderly phase-out plan. Premature release of the information among contractor employees could have had a serious effect on the operational posture. Realizing the short-term prospects of the OXCART program, technicians might seek employment elsewhere.

Contract management recognized the delicacy of the situation and by careful planning, could maintain their capabilities and program excess personnel into other activities at the appropriate time.

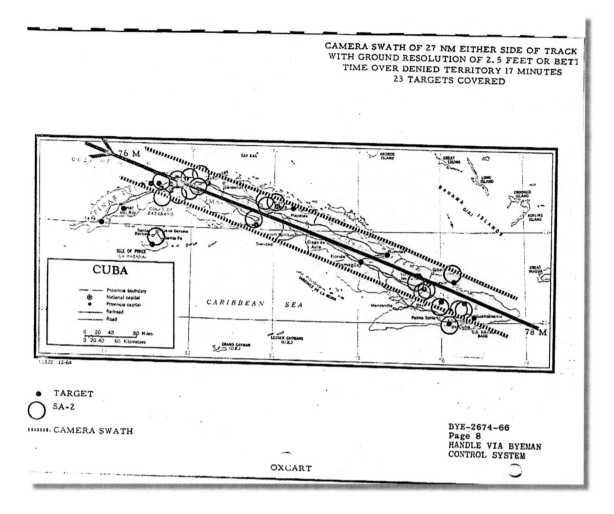

With the program committed to a specific timetable, project headquarters took certain specific actions. It removed systems in, or scheduled for flight test, unless very near operational acceptance, and placed them in storage or made them available to other National Reconnaissance Office programs. These included second-generation ECM systems, the side looking radar, the infrared sensor, and the system ELINT collection gear. The CIA adopted an aircraft standard operational ready configuration and removed from stock and returned to depot storage any types of equipment surplus to the standard configuration. It withdrew from the inventory and stored two of the Perkin Elmer cameras scheduled for overhaul and update.

On 16 May 1967, the president directed the deployment of OXCART to Kadena. Okinawa, for reconnaissance of North Vietnam for the primary purpose of detecting, surprise introduction of offensive missiles. The CIA deployed three aircraft and flew the first mission on 31 May 1967.

The deployment had the effect of slowing actions to phase out OXCART, but the 31 December 1967 termination date still existed at the time of the deployment. In June, the CIA proposed two options to modify the phase-out given the Kadena implementation. One was to defer mothballing of any aircraft until 31 December 1967 to ensure a backup capability for the deployed aircraft. The second option was to continue the entire OXCART fleet through FY 1968.

The director, National Reconnaissance Office was amenable to the first alternative, influenced by the fact that the SR-71 had not yet achieved an operational ECM capability. October was the earliest date it would be ready.

The deputy secretary of defense modified the delay in mothballing aircraft by recommending putting one test aircraft down in July and the remaining eight in December 1967. He further directed that Strategic Air Command be responsible for Kadena operations in December 1967 using its SR-71s.

At the 12 September 1967 meeting of the NRP Executive Committee Executive Committee, Admiral

Taylor stated that the CIA would withdraw the OXCART aircraft from Kadena by mid-November to make way for the SR-71 operation.

He pointed out the reassignment of experienced contractor personnel from the program beginning 1 December, and the depletion of the spare parts supply to a point where it would be most difficult to sustain an operational posture. He emphasized the October deadline as the effective date of any meaningful change in the phase-out schedule. Dr. Hornig stressed the importance of having the SR-71 operationally ready before reaching the point of no return in the phase-out of the OXCART. He requested a current comparison of the OXCART and SR-71, which Dr. Flax, the Director, National Reconnaissance Office, said he would provide for review by the Executive Committee before 1 October.

The minutes of the Executive Committee meeting of 29 September, which best reflected the situation vis a vis the SR-71 deployment for BLACK SHIELD. In them, Mr. Nitze stated that the purpose of this meeting was to review the status of the SR-71, to determine what problems may have arisen from SR-71 Category II Tests, and to recommend actions for the future on OXCART phase down and SR-71 deployment. Dr. Flax referred briefly to the papers he had distributed to the members, pointing out that they summarized about two-thirds of the information available from the SR-71: Category H Tests officially concluding in two days on October 1, 1968. He stated that, in general, the SR-71 was in a satisfactory state, and that operational experts judged that the air force could assume the North Vietnam missions on December 1, 1967. A joint chiefs of staff statement reflected this view in that the SR-71 was ready for operational employment.

Dr. Hornig referred to the documents furnished to the Executive Committee and data dealing with the vulnerability of both the OXCART and SR-71 aircraft. He pointed out that, from his assessment of the data, and based on the listed equipment statistical factors, and performance curves, the SR-71 appeared to be two to four times more vulnerable than the OXCART.

There followed a detailed discussion on vulnerability studies, operational techniques and impact, ECM systems and capabilities, the present activity of the enemy, his intentions in the future, and the outlook for future operations. Hornig then compared the payload volumes of the aircraft and the photographic swath widths of their sensors.

He believed the committee should not be too hasty in deciding to deploy the SR-71. Dr. Flax stated that a simple comparison of sensor swath widths was not, in his view, a logical way to compare the mission coverage capabilities of the aircraft. He assumed a factor of two invulnerability on this basis did not reflect mission requirements in any event since complete area coverage of North Vietnam was not being sought or achieved. Mr. Nitze outlined the following options for consideration:

(1) Delay the transition from OXCART to the SR-71,
(2) Recommend a reversal of the December 1966 decision, or
(3) Adhere to that decision.

The discussion turned to the first option. Asked the desirability of this option, Dr. Flax stated that if there were no economic restraints whatever, he preferred to retain the total force. However, financial constraints were genuine, and he believed it called for a firm decision now. He pointed that a six-month delay in making the transition could cost $32. 0 million.

To a question raised as to the cost of reclaiming OXCART aircraft from storage if required, he stated that this would involve approximately $300 to 500 thousand per aircraft if done within the first year. He also pointed out that the present financial plan provided some OXCART overlap by providing for continued operation at Area 51 during December. Mr. Nitze observed that the additional $32.0 million required for a delay in transition from the OXCART to the SR-71 was extremely critical in today's budgetary environment. Dr. Hornig favored a delay of six months.

Dr. Foster agreed with the concept of delay but recommended a shorter period somewhere between three and six months. Mr. Hoffman stated that they should follow unchanged the original decision of last December. Mr. Helms favored a delay, which Dr. Flax agreed would provide more confidence in assuring continued operational effectiveness in the face of possible improvement of North Vietnamese defenses.

Vietnam was just now beginning to be brought to bear on the OXCART aircraft.

The cost of this insurance would, of course, relate to the length of delay in phase out. He believed they should not hold the SR-71 deployment until the very last day of any agreed to delay period if recommended a three-month delay. They should schedule the SR-71 deployment for mid-February 1968. "Mr. Nitze asked for a memorandum spelling out

(1) What the decision for a three-month delay would do for the program

(2) The associated costs, and

(3) What they intended for the interim?

After Dr. Flax had prepared such a paper, Mr. Nitze stated he would confer with the secretary of defense on this matter early in the following week.

The deliberations which followed resulted in a decision to retain the three OXCART aircraft at Kadena through 1 February 1968. Strategic Air Command would assume BLACK SHIELD operations by 15 February. All operational aircraft would remain flying until mothballed on 31 March 1968. (A Director, National Reconnaissance Office memorandum of 3 October 1967 to the deputy secretary of defense contained the details of the extension and the revised phase-out plan.)

On 29 December 1967, the deputy secretary of defense advised the director, National Reconnaissance Office of the extension of the OXCART program through 30 June 1968.

The SR-71 was to assume North Vietnamese reconnaissance by 15 March 1968, and the OXCART aircraft to remain at Kadena for 30 days after that to provide a contingency overlap.

The original decision to terminate the program, made in December 1966, tended now to be obscured by the BLACK SHIELD deployment and the two extensions of the termination date. That decision now appeared open to question, and the feeling prevailed that Presidential reaffirmation was required.

The president's special assistant, Mr. Walt Whitman Rostow, had expressed the opinion that they should not undertake the removal of the OXCART capability of the Far East unless the president specifically approved. Key member congressional figures and members of the president's Foreign Intelligence Advisory Board and the president's Scientific Advisory Committee were also concerned with the loss of OXCART. Consequently, the Executive Committee decided to reopen the matter and once again examine alternatives to outright phase-out of the program. Again, the arguments marshaled in defense of retaining the OXCART capability, buttressed now by its demonstrated operational performance. In March, the SR-71 unit deployed to Kadena with three aircraft and on 15 March assumed the BLACK SHIELD mission.

The OXCART aircraft remained in place to provide a contingency backup capability, and to provide coverage of North Korea because of the Pueblo affair. A study of the feasibility and cost of continuing the OXCART program beyond its phase-out date completed in the spring of 1968 by the director, National Reconnaissance Office with four alternatives considered:

1. Transfer all OXCART aircraft to Strategic Air Command by 31 October 1968 substitute air force for contractor support where possible turn the test A-12 aircraft over to the SR-71 test facility. FY 1969 costs would be 62, 160,000.

2. Transfer OXCART as in Alternative 1, above, and store eight SR-71s. FY 1969 costs will be $40,960,000.

3. Close Area 51 and collocate OXCART fleet with SR-71s at Beale Air Force Base, California, but with CIA retaining control and management. The FY 1969 cost will be $72,240, 000. 4. Continue OXCART operations at Area 51 under CIA control and management. FY 1969 costs will be 72,000,000.

Mr. Helms expressed his reactions to the alternatives considered by the National Reconnaissance Office in a Memorandum to Messrs. Nitze, Hornig, and Flax dated 18 April. 1968. In it, he questioned why, if they could store eight SR-71s in one option, they could not keep them in all the options, and have the resultant savings applied in each case. He questioned the lower, cost figures of combining the A-12s with the SR-71s and disagreed, for security reasons, with co-locating the two fleets. Above, however, he felt the key question was the desirability of retaining a covert civilian capability. He judged that such a requirement continued. In conclusion, he flatly recommended maintenance of the OXCART capability at Area 51 under

CIA management. Time, however, was fast running out of the OXCART program.

The prime consideration in every phase of discussions concerning the OXCART was budgetary. They made very clear in the penultimate paragraph of the minutes of the Executive Committee meeting of 29 April 1968. Despite the very cogent reasoning of the Director, Central Intelligence, supported by the President's Scientific Adviser, the deputy secretary of defense felt the central issue was budgetary since the FY 1969 budget assumed termination of the OXCART program for the lack of money available to sustain the program.

On 16 May 1968, the: secretary of defense reaffirmed the original decision to terminate the OXCART program and store the aircraft. At his weekly luncheon with his principal advisers on 21 May 1968, the president confirmed the secretary's decision.

Edwards AFB Takes Over - Commanded by Lt Col Larry D. McClain.

Sampson had transferred to OSA headquarters with Herb Saunders in 1964 for assignment to the OXCART program. At the time, the CIA had 12 security officers assigned to handle IDEAL-IST and OXCART. Sampson wrote the first industrial security manual for special projects, his manual becoming the forerunner to the current manual at Area 51. In those days, the CIA security, contracts, and technology traveled as a team, all decision makers, and 15 considered a crown as compared to 100 to 250 today.

At his suggestion, Col Jack Ledford transferred him to Los Angeles to open an office to cut down on travel and to speak with one voice to the contractors. He merged the contracting officers with Col Ralph Ford and Maj Harvey Cohen, the Air Force Security representatives, to teach them the security system desired by the CIA. At the time, Norm Nelson was still operating out of the filthy, 10 x 10-foot closet in a Lockheed hangar in Burbank.

Their primary job as security was making sure that no unauthorized person-boarded the White Whale transporting workers to and from Area 51, and to monitor Lockheed's security performance. The White Whale consisted of three Lockheed Constellations.

Sampson persuaded Kelly Johnson to provide suitable office space in the Skunk Works, and he made available a 10 x 15-foot room next to the CIA Commo Center. Sampson continued seeking an office since he did not want Lockheed knowing the CIA's other business.

Sampson managed to establish the Western Industrial Liaison Detachment (WILD) in the basement of the Tishman building near LAX. He arranged for Pratt & Whitney to dispatch their corporate interior decorator to Los Angeles to design the interior with furniture and pictures. A longtime supporter of the CIA's programs employed by Hughes Aircraft arranged for Hughes guards to patrol the CIA's space every two hours during non-working hours.

It was not long before a communicator, auditors, and finance personnel descended upon his office once the others at Langley realized that he now had office space. From that point on, his office had some handling satellites and this group working the aviation business. Their venue became everything west of the Mississippi.

Dick Sampson enjoyed telling a Kelly Johnson story. "Lockheed had over 100 disapproved employees working on the programs. Unless gay or communists, Kelly would not terminate them. He hired any warm body with the necessary skills. Lockheed hired them, and the next day they reported to work in the Skunk Works."

Kelly and Sampson engaged in several heated debates over this situation. Johnson usually got OSA headquarters to overrule Sampson. To keep this situation from proliferating, Sampson had headquarters conduct a study to determine the reasons for disapproval. Fifty-two percent of the time, it was either most recent employment, or police, and credit.

Sampson finally arranged for the SAC of an OS investigative office under commercial cover in Los Angeles to conduct a mini-investigation within 48 hours to make sure there was no disqualifying information.

Armed with these statistics, Johnson reluctantly agreed not to put anyone to work at the Skunk Works for 48 hours. If the subject failed to pass the initial screening, he would not put him or her work.

Within a year, the WILD office came to represent the DDS&T. Sampson's group met foreign and other dignitaries of interest to DDS&T.

Epilogue

It did not happen overnight, but Gen Curtis LeMay did what he said he would before the U-2. Once the CIA got the U-2 and A-12 programs running and paid for, he took the programs away for his air force. However, he could not take away the CIA's accomplishments of pioneering Area 51, the U-2, the A-12 that brought new technology with exotic materials and stealth.

The CIA had left Area 51 once, after the U-2, but wasn't leaving again just because it lost the OXCART program. The CIA was now looking at satellite reconnaissance, the Aquiline project, a stealthy propeller-driven, low altitude, anhedral-tailed UAV, and others too highly classified to mention. The CIA had something that the air force did not. The CIA had a new division called Science and Technology, thanks to Area 51. The CIA also had its hand-picked specialists embedded in EG&G project. It had a new business also called science and technology. Now it needed customers for its business.

Even as the OXCART program was shutting down, the CIA welcomed its first customers whom all came to Area 51 with a common problem — the Soviet MiGs out fighting them in Vietnam with a 9:1 kill ratio against the US Air Force and Navy. The first customers were:

- AFSC (Air Force Systems Command),
- The Laboratories at Wright-Patterson AFB,
- The AFFTC (Air Force Flight Test Center),
- SAC (Strategic Air Command),
- NDA (National Defense Agency),
- NASIC (National Air and Space Intelligence Center),
- NATC (Naval Air Test Center),
- The NWC (Naval Weapons Center),
- The TAC (Tactical Air Command), and
- CIA special projects contractor specialists and their facilities at Area 51. Nevada.

As the Blackbirds moved out, the MiG-21 Fishbed and the air force and navy bandits moved in for Project DOUGHNUT.

In January 1968, Project HAVE DOUGHNUT, a joint USAF/Navy technical and tactical evaluation of the MiG-21F-13 began at Area 51 at the same time the Blackbird family moved out. Thus, began an era in Area 51 history where once a project concluded, the client moved out, and another moved in. Werner Weiss, the CIA's commander at Area 51, handed the reins over to Richard A. "Dick" Sampson, who in turn did the same for Sam Mitchel. It was lights out for the CIA at Area 51 when Mitchell transferred to the control of the United States Air Force Detachment 3, AFFTC at Edwards commanded by Lt Col Larry D. McClain in April 1979.

Glossary

AEC   The United States Atomic Energy Commission (AEC) was an agency of the United States government established after World War II by Congress to foster and control the peacetime development of atomic science and technology. President Harry S. Truman signed the McMahon/Atomic Energy Act on August 1, 1946, transferring the control of atomic energy from military to civilian hands, effective from January 1, 1947. The Energy Reorganization Act of 1974 abolished the agency and assigned its functions to two new agencies: The Energy Research and Development Administration and the Nuclear Regulatory.

CAT	Civil Air Transport
Comp	Abbreviation for "complimentary
COMSEC	Communications Security
CONUS	Continental United States
CRYPTO	Cryptographic
CW	Continuous Wave
CWAR	Continuous Wave Acquisition Radar
DPRK	North Korea

DREAMLAND - Name of the Groom Lake facility in Nevada during the USAF stealth Project HAVE

DTS	Data Transmission System
DYCOMS	Dynamic Coherent Measurement System used for RCS
ECCM	Electronic Counter-Countermeasures
ECM	Electronic Counter Measurers
EG&G	Edgerton, Germeshausen, and Grier, Inc. EG&G is a United States national defense contractor and provider of management and technical services.
EMT	Emergency Medical Technician

ESPIONAGE – a covert act of spying on others often a systematic use of spies by a government to discover Military, Political, Economic secrets

FAA	Federal Aviation Administration
FCF	Functional Check Flight
FOUO	For Official Use Only
IFF	Identification Friend or Foe
Isinglass	CIA/General Dynamics Mach 4/5 rocket-powered, air-launched reconnaissance vehicle study to replace Oxcart, capable of reaching orbital velocities, not completed/never flown
JAG	Judge Advocate General's Corps
LVCP	Landing Craft, Vehicle, Personnel. A Higgins boat
MACV	Military Assistance Command Vietnam
MOS	Military Occupational Specialty
NASA	National Aeronautics and Space Administration
NASA	High Range A 400-mile high speed corridor extending from NASA/Dryden at Edwards AFB to Ely, Nevada with Beatty tracking station being in the middle of the corridor.
NERVA	Nuclear Engine for Rocket Vehicle Application
Nevada Proving Ground	The site, established on 11 January 1951, for the testing of nuclear devices
NOFORN	Not Releasable to Foreign Nationals (access restricted to US citizens)
NRDS	Nuclear Rocket Development Station located at Jackass Flats within the AEC's atomic testing grounds
OCS	Officer Candidate School

Operation BLACK SHIELD The operational phase of Project Oxcart flying the A-12 over North Vietnam and North Korea

PARADISE RANCH Name of the Groom Lake facility in Nevada during the CIA A-12 Project OXCART

Project Aquatone	Initial program name for CIA-sponsored U-2 Reconnaissance plane
Project Aurora	USAF, classified program, most likely for B-2A procurement
Project Bald Eagle	USAF counterpart to CIA's Aquatone; big-wing B-57 Canberra, became RB-57D
Project Gusto	A 1957 advisory committee selected by the CIA to select a successor to the U-2
Project HAVE BLUE	The Lockheed XST ultrasecret stealth prototype that developed into the F-117 Nighthawk
Project HAVE DOUGHNUT	One MiG-21F-13, used for Air Combat Training at Groom Lake, USAF/USN joint project, predating Have Idea (1968)
Project HAVE DRILL	2 ex-Syrian MiG-17F from Israel, used for Air Combat Training at Groom Lake, USAF/USN joint project, predating Have Idea (1969)
Project HAVE FERRY	Exploitation of a MiG 17.
Project HAVE GARDEN	Evaluation of Soviet R-13-300 turbojet engine (used in MiG-21) blade profiling and compressor maps by Pratt & Whitney (1978)
Project HAVE IDEA	Air Combat Training with various MiGs, to Constant Peg
Project HAVE RADIO	Exploitation of a foreign radar system.
Project OXCART:	The CIA project at Groom Lake to design, build, and operate the A-12 Blackbird reconnaissance plane to replace the U-2. Originally classified Top Secret, OXCART was UNCLASSIFIED according to Senior Crown Security Class Guide dated 11/01/89, approved and dated 25 February 91. The CIA declassified the identity of the personnel in September 2007, and the restriction on using the name Area 51 lifted September 2010.
PX	Post Exchange
Pylon	The Special Projects team preferred to call the pylon a pole as pylon also identified the tie point under aircraft wings where they fastened external loads.
RatScat	Radar Target Scatter Site
Reconnaissance	An overt act of reconnaissance in the field. A search made to produce useful military information to inspect, observe, or survey for enemy positions, strengths, and intent, etc. No longer just military purposes in this blended Technology Age
ROK	South Korea
SCR	Short for Signal Corps Radio # 584
SPYING -	A person employed by a Gov to obtain secret info or Intel on another Gov Alt – any person who clandestinely seeks info on people or projects for a profit
Tank Battery	An installation of identical or nearly identical oil storage tanks
TO&E	Table of Organization and Equipment
TOC	Tactical Operations Center
UCMJ	Uniform Code of Military Justice
USO	United Service Organizations
USOM	Office of Rural Affairs (Counterinsurgency)
USS Pueblo	American spy ship captured by the North Korean navy in February 1968
WATERTOWN –	Name of the Groom Lake facility in Nevada during the CIA U-2 Project AQUATONE
WNINTEL	Warning Notice - Intelligence Sources and Methods Involved
X-15	The North American X-15 rocket-powered aircraft/spaceplane member of the X-series of experimental aircraft
ZULU	Zulu Time is the world time used by international shortwave broadcasters, ham radio operators, shortwave listeners, the military, plane and ship navigation, and utility radio services — also known as UT or UTC (Universal Time [Coordinated)). The entire planet is on the same time. There are no time zones for UTC. UTC also has no Daylight-Saving Time or Summer Time. 0800 hours at the Beatty Tracking station was 1600 hours ZULU time.

Bibliography

Bissell, Richard M., Jr., with Jonathan E. Lewis and Frances T. Pudlo. Reflections of a Cold Warrior: From Yalta to the Bay of Pigs. New Haven, CT: Yale University Press, 1996.

Brown, William H. "J58/SR-71 Propulsion Integration." Center for the Study of Intelligence 26, no. 2 (Summer 1982): 15–23.

Central Intelligence Agency, Office of Special Activities. "Chronology, 1954–68." Declassified, June 2003. The CIA and the U-2 Program 1954–1974 by Gregory W. Pedlow and Donald E. Weizenbach, History Staff, Center for the Study of Intelligence, CIA, 1998

CIA BYE 2986–65. Project Oxcart and Operation BLACK SHIELD Briefing Notes. October 20, 1965.

Crickmore, Paul F. Lockheed SR-71: The Secret Missions Exposed. London: Osprey, 1996.

Drendel, Lou. SR-71 Blackbird in Action. Carrollton, TX: Squadron/Signal Publications, 1982.

Goodall, James. SR-71 Blackbird. Carrollton, TX: Squadron/Signal Publications, 1995.

Goodall, James, and Jay Miller. Lockheed's SR-71 "Blackbird" Family: A-12, F-12, M-21, D-21, SR-71. Hinckley, UK: Midland Publishing, 2002.

Graham, Richard H. SR-71 Blackbird: Stories, Tales, and Legends. St. Paul, MN: Zenith Press, 2002.

———. SR-71 Reveals: The Inside Story. Osceola, WI: Motorbooks International Publishing, 1996.

Haines, Gerald K. "The CIA's Role in the Study of UFOs, 1947–90." Center for the Study of Intelligence 41, no. 1 (1997): 67–84.

Helms, Richard, with William Hood. A Look Over My Shoulder: A Life in the Central Intelligence Agency. New York: Random House, 2003.

Jenkins, Dennis R. Lockheed SR-71/YF-12 Blackbirds. North Branch, MN: Specialty Press, 1997.

Johnson, Clarence L. "Development of the Lockheed SR-71 Blackbird." Center for the Study of Intelligence 26, no. 2 (Summer 1982): 3–14.

Johnson, Clarence L. "Kelly," with Maggie Smith. Kelly: More Than My Share of It All. Washington, D.C.: Smithsonian Institution Press, 1985.

Landis, Tony R. Lockheed Blackbird Family: A-12, YF-12, D-21/M-21 and SR-71 Photo Scrapbook. North Branch, MN: Specialty Press, 2010.

McIninch, Thomas P. "The OXCART Story." Center for the Study of Intelligence 15, no. 1 (Winter 1971): 1–25.

Merlin, Peter. From Archangel to Senior Crown: Design and Development of the Blackbird. Reston, VA: American Institute of Aeronautics and Astronautics, 2008.

Miller, Jay. Lockheed Martin's Skunk Works. Leicester, UK: Midland Publishing, 1995.

Pedlow, Gregory W., and Donald E. Welzenbach. The Central Intelligence Agency and Overhead Reconnaissance: The U-2 and OXCART Programs, 1954–1974. Washington, D.C.: Central Intelligence Agency, 1992. Chapter 6 on OXCART declassified October 2004.

Remak, Jeannette, and Joseph Ventolo, Jr. A-12 Blackbird Declassified. St. Paul, MN: MBI Publishing Co., 2001.

———. The Archangel and the OXCART: The Lockheed A-12 Blackbirds and the Dawning of Mach III Reconnaissance. Bloomington, IN: Trafford Publishing, Co., 2008.

Rich, Ben R., and Leo Janos. Skunk Works: A Personal Memoir of My Years at Lockheed. Boston: Little, Brown, 1994.

Richelson, Jeffrey T. The Wizards of Langley: Inside the CIA's Directorate of Science and Technology. Boulder, CO: Westview Press, 2001.

Robarge, David, PhD. Archangel: CIA's Supersonic A-12 Reconnaissance Aircraft. 2nd ed. Washington, D.C.: Center for the Study of Intelligence, Central Intelligence Agency, Government Printing Office, January 2012.

Suhler, Paul A. From Rainbow to GUSTO: Stealth and the Design of the Lockheed Blackbird. Reston, VA: American Institute of Aeronautics and Astronautics, 2009.

Sweetman, Bill. Lockheed Stealth. St. Paul, MN: MBI Publishing, 2001.

Wheelon, Albert D. "And the Truth Shall Keep You Free: Recollections by the First Deputy Director of Science and Technology." Center for the Study of Intelligence 39, no. 1 (Spring 1995): 73–78.

Whittenbury, John R. "From Archangel to OXCART: Design Evolution of the Lockheed A- 12, First of the Blackbirds." PowerPoint presentation, August 2007.

Wings of Fame. London: Aerospace Publishing, 1997.

Reference Documents

"History of the OXCART Program." Burbank, CA: Lockheed Aircraft Corporation, July 1, 1968. Declassified, August 2007.

Johnson, Clarence L. "Archangel Log." Undated.

Websites

archive.org/stream/HistoryOfTheOfficeOfSpecialActivitiesFromInceptionTo1

area51specialprojects.com

blackbirds.net

roadrunnerinternationale.com

www.habu.org

www.Lockheedmartin.com

About the Author

Thornton D. "TD" Barnes, author, and entrepreneur, grew up on a ranch at Dalhart, Texas. He graduated from Mountain View High School in Oklahoma and embarked on a ten-year military career. He served as an Army intelligence specialist in Korea and then continued his education while in the US Army, attending two and a half years of missile and radar electronics by day and college courses at night. Barnes deployed with the first combat Hawk missile battalion during the Soviet Iron Curtain threat before attending the Artillery Officer Candidate School, where an injury ended his military career.

Barnes' career includes serving as a field engineer at the NASA High Range in Nevada for the X-15, XB-70, lifting bodies and lunar landing vehicles; working on the NERVA project at Jackass Flats, Nevada; and serving in Special Projects at Area 51. Barnes later formed a family oil and gas exploration company, drilling, and producing oil and gas and mining uranium and gold.

Barnes currently serves as the CEO of Startel, Inc., a landowner, and is actively mining landscape rock and gold in Nevada. He serves as the president of Roadrunners Internationale, an association of Area 51 veterans, and is the executive director of the Nevada Aerospace Hall of Fame.

Two National Geographic Channel documentaries feature Barnes: Area 51 Declassified and CIA—Secrets of Area 51. Numerous documentaries on the History Channel, the Discovery Channel, the Travel Channel, and others also feature him. The Annie Jacobsen book Area 51 Declassified documents his career.

Barnes lives in Henderson, Nevada.

Connect with the Author Online
Facebook: www.facebook.com/ThorntondBarnes
Blog: td-barnes.com/blog
Website: td-barnes.com
LinkedIn: www.LinkedIn.com/profile/edit?trk=tab_pro
Twitter: twitter.com/ThorntonDBarnes

Other Books by the Author

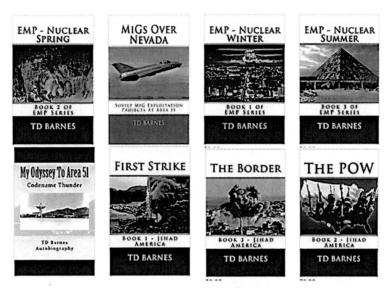

Fiction
EMP - Book 1 - Nuclear Winter
EMP - Book 2 - Nuclear Spring
EMP - Book 3 - Nuclear Summer
American Jihad - Book 1 - First Strike
American Jihad - Book 2 - The POW
American Jihad - Book 3 - The Border

Non-fiction
My Odyssey to Area 51
MiGs Over Nevada

THE AREA 51 CHRONICLES the CIA AT AREA 51 1955–1979
　　　　　　Book 1 - The Angels
　　　　　Book 2 - The Archangels
　　　　Book 3 - The Company Business
Released September 4, 2017
The Secret Genesis of Area 51

CPSIA information can be obtained
at www.ICGtesting.com
Printed in the USA
LVOW09s1616211217
560228LV00049B/3388/P

9 781547 084876